PROPERTY

COMPASSION IN W
INFORMATION LIBRARY

PETERSFIELD GU32 3EH

PLEASE BOOK OUT BEFORE BORROWING

STEREOTYPIC ANIMAL BEHAVIOUR

Dedicated to David Wood-Gush for his
contributions to the study of farm animal
stereotypies and for his friendship.

STEREOTYPIC ANIMAL BEHAVIOUR
Fundamentals and Applications to Welfare

Edited by

ALISTAIR B. LAWRENCE

Genetics and Behavioural Sciences Department
The Scottish Agricultural College
Edinburgh, UK

and

JEFFREY RUSHEN

Agriculture Canada Research Station
Lennoxville, Quebec
Canada

CAB INTERNATIONAL

CAB INTERNATIONAL
Wallingford
Oxon OX10 8DE
UK

Tel: Wallingford (0491) 832111
Telex: 847964 (COMAGG G)
Telecom Gold/Dialcom: 84: CAU001
Fax: (0491) 833508

© CAB INTERNATIONAL 1993. All rights reserved. No part of this publication may be reproduced in any form or by any means, electronically, mechanically, by photocopying, recording or otherwise, without the prior permission of the copyright owners.

A catalogue record for this book is available from the British Library.

ISBN 0 85198 824 5

Typeset by BP Integraphics, Bath
Printed and bound in the UK by Redwood Books, Trowbridge

Contents

Contributors		vii
Foreword PIET WIEPKEMA		ix
1.	Introduction ALISTAIR B. LAWRENCE AND JEFFREY RUSHEN	1
2.	Forms of Stereotypic Behaviour GEORGIA J. MASON	7
3.	The Motivational Basis of Stereotypies JEFFREY RUSHEN, ALISTAIR B. LAWRENCE AND E.M. CLAUDIA TERLOUW	41
4.	The Concept of Animal Boredom and its Relationship to Stereotyped Behaviour FRANÇOISE WEMELSFELDER	65
5.	Stress and the Physiological Correlates of Stereotypic Behaviour JAN LADEWIG, ANNE MARIE DE PASSILLÉ, JEFFREY RUSHEN, WILLEM SCHOUTEN, E.M. CLAUDIA TERLOUW AND EBERHARD VON BORELL	97
6.	Neurobiological Basis of Stereotypies SIMONA CABIB	119

7.	Functional Consequences of Behavioural Stereotypy ROBERT DANTZER AND GUY MITTLEMAN	147
8.	Future Research Directions FRANK O. ÖDBERG	173
9.	Conclusions and Implications for Animal Welfare IAN J.H. DUNCAN, JEFFREY RUSHEN AND ALISTAIR B. LAWRENCE	193
Index		207

Contributors

SIMONA CABIB, *Instituto di Psicobiologia e Psicofarmacologia (CNR), via Reno 1, Roma 1–00198, Italy.*

ROBERT DANTZER, *INRA-INSERM U176, Rue Camille Saint-Saens, 33077 Bordeaux Cedex, France.*

IAN J.H. DUNCAN, *Department of Animal and Poultry Science, Ontario Agricultural College, University of Guelph, Guelph, Ontario, Canada N1G 2W1.*

JAN LADEWIG, *Institut fur Tierzucht und Tierverhalten, FAL, Trenthorst, 2601 Westerau, Germany.*

ALISTAIR B. LAWRENCE, *Genetics and Behavioural Sciences Department, The Scottish Agricultural College Edinburgh, West Mains Road, Edinburgh, EH9 3JG, UK.*

GEORGIA J. MASON, *Sub-Department of Animal Behaviour, University of Cambridge, Madingley, Cambridge, CB3 8AA, UK.*

GUY MITTLEMAN, *Department of Psychology, Memphis State University, Memphis, TN 38152, USA.*

FRANK O. ÖDBERG, *Faculty of Veterinary Medicine, State University of Ghent, Heidestraat 19, B-9220 Merelbeke, Belgium.*

ANNE MARIE DE PASSILLÉ, *Agriculture Canada Research Station, Lennoxville, Quebec, Canada J1M 1Z3.*

JEFFREY RUSHEN, *Agriculture Canada Research Station, Lennoxville, Quebec, Canada J1M 1Z3.*

WILLEM SCHOUTEN, *Department of Animal Husbandry, Wageningen Agricultural University, Marykeweg 40, Postbox 338, 6700 AH Wageningen, The Netherlands.*

E.M. CLAUDIA TERLOUW, *Genetics and Behavioural Sciences Department, The Scottish Agricultural College Edinburgh, West Mains Road, Edinburgh, EH9 3JG, UK.*

EBERHARD VON BORELL, *Department of Animal Science, Iowa State University, 337 Kildee Hall, Ames, IA 50011, USA.*

FRANÇOISE WEMELSFELDER, *Institute of Theoretical Biology, Rijksuniversiteit te Leiden, Kaiserstraat 63, Postbus 9516, 2300 RA Leiden, The Netherlands.*

Foreword

When the editors asked me to write a preface to their forthcoming book on stereotypic animal behaviour, I accepted spontaneously. I did so because this topic is of utmost relevance to present day discussions about animal welfare, and secondly, this theme has strong connections with brain-behaviour problems that are near to my heart. Moreover, writing a preface seems to be a respected activity. I did not envisage, however, that such an excellent manuscript was in the making as I have on my desk now; a well-integrated book of reference on stereotypic behaviour in vertebrates.

Stereotypic behaviour, stereotypies, are typical of chronically disturbed vertebrates. The occurrence of these stereotypies refers to serious shortcomings in the environment of their performers, including their housing during development, although the effect of environment is also strongly influenced by the coping capabilities of the individuals involved. Stereotypies may also be elicited by specific pharmacological brain manipulations. Stereotypies are characteristic of humans and other vertebrates when they lose control over their actual life conditions.

In order to prevent or to reduce these negative behavioural symptoms we have to understand their causation and function, at least from a biological point of view. Insight into the processes or mechanisms that underlie stereotypies in the individual or species is of great pure and applied value. Among individuals the occurrence of stereotypies varies according to differences in age, sex, experience and coping style; while between species large differences exist according to their different ecological adaptations. The applied aspects derived from present research on stereotypies concern the development of adequate medical care in humans and appropriate designs in animal husbandry.

The great value of this book is its being written by authors who are all active and leading experts in the field of stereotypic animal behaviour: ethologists, neurobiologists, pharmacologists. They all present their data and thoughts in a well-integrated manner. This makes the book not only very instructive, but also attractive. Although the varying methods and models have different flavours, they do not lead to contradictory explanations. An interesting example of divergent theorizing is found in the question as to whether animals suffer when performing stereotypies.

Most chapters of this book illustrate that the original pure pharmacological approach to understanding stereotypic behaviour and the broader biological one studying the 'natural' occurrence of stereotypies contact each other and are on the verge of a useful integration. From the point of view of brain-behaviour relationships this is a substantial step forward. Obviously the research on stereotypic behaviour in animals has reached a stage where we have a much better idea of the sorts of questions we need to answer to understand and control this behaviour. Since the book is a clear representation of this stage of development, it will be of great significance for those interested in behavioural biology and its application to animal husbandry. Moreover, relevant information is offered for psychiatrists and neurologists trying to understand deviant behaviour in primates and man.

Stereotypic Animal Behaviour is an exciting piece of work.

Piet Wiepkema

Introduction 1

ALISTAIR B. LAWRENCE[1] AND JEFFREY RUSHEN[2]
[1]*The Scottish Agricultural College, Edinburgh, UK;* [2]*Agriculture Canada Research Station, Lennoxville, Quebec, Canada.*

Over the millennia, we have confined a large range of animal species, and, in almost all cases, this confinement has brought us some advantage. For example, confining meat-producing animals gives better control over their reproduction and nutrition thus increasing their productivity. Confining wild animals allows us to observe them at our leisure and in safety. Confinement also has some advantages for the animal, such as an improved food supply, decreased risk of predation and protection from environmental destruction. It is, however, only relatively recently that there has been a widespread appreciation that confinement may also have undesirable effects on animals. There has been concern over housing of zoo animals since the 1950s (e.g. Hediger, 1955); concern over confinement of farm animals was stimulated by Ruth Harrison's (1964) book *Animal Machines: The New Factory Farming Industry*, which graphically described the restrictive environments developed for modern farm livestock. Harrison's book had a strong impact on a public that appeared largely unaware of the changes that had occurred in farming practice as a result of the post-war pursuit of 'cheap food'. Similar public appreciation of animal welfare issues was being expressed in other European countries. Since that time, public concern over the 'welfare' of confined animals has increased and now extends well beyond zoo and farm animals to include laboratory and companion animals.

The increased public concern caused scientists to become more involved in the 'welfare' debate. In the United Kingdom, the Brambell Committee was established in response to *Animal Machines* to review evidence that confinement of farm animals causes undue 'suffering'. Much of the evidence given to the committee was delivered by behavioural scientists, establishing a link between the scientific study of behaviour and the assessment of animal 'welfare' that has remained over the years. An important aspect of scientific debate was the effect that confinement has on animal behaviour. In his appendix to the Committee's report, the eminent ethologist W. H. Thorpe wrote 'whilst accepting the need for much restriction, we must draw the line at conditions which completely suppress all or nearly all the natural instinctive urges and behaviour patterns...' (Command Paper 2386, 1965). Elsewhere in the report (p. 15) mention is made of the undesirable habits and 'vices' that develop under intensive conditions. One form of 'undesirable habit' which has maintained public and scientific interest since that time is so-called 'stereotypic' behaviour patterns.

The pacing or swimming patterns of zoo-kept bears, the 'bar-biting' of closely confined sows, the locomotor 'loops' and 'somersaults' of caged mink and voles, and the 'crib-biting' and 'wind-sucking' of stalled horses are well-known examples of stereotypies. But there are a plethora of other behaviours, performed in closely confining and barren environments, or under supposedly stressful conditions or induced by drugs, that are similarly labelled as stereotypic behaviours (Chapter 2). Despite their heterogeneity, all these behaviours were certainly first noticed because of their 'abnormal' appearance, and despite attempts to objectively define stereotypies (Chapter 2) many scientists would probably still subjectively describe stereotypies as being qualitatively different or even 'bizarre' relative to the normal flow of behaviour.

Given the range of stereotypic behaviours, it is not at all surprising that they are studied by scientists for a number of different purposes. Stereotypies induced by drugs such as amphetamine have been studied as a model for drug abuse and the neural basis of certain brain or psychiatric disorders in humans (Chapter 6). Stereotypies (or adjunctive behaviours) induced by intermittent food reward have been used as models of 'stress-induced' behaviour and studied with respect to their function as coping behaviours (Chapter 7). Most central to this book has been the study of the stereotypies induced by confining environments such as in farm, laboratory and zoo animals (Chapters 3, 4 and 5). Here the study of stereotypies has had a more applied aspect being linked to the concern over animal 'welfare', particularly in farm animals. There is circumstantial evidence of a link between the development of stereotypies in farm animals and the apparently 'stressful' or 'frustrating' circumstances under which they develop (e.g. Chapter 5). Stereotypy research in farm animals has been broadly directed towards the development of stereotypies as indicators of 'stress', 'suffering' and hence reduced 'welfare'.

Assuming that the scientific interest in environmental stereotypies, at least in farm animals, would have been stimulated by the Brambell Committee, one can ask what progress has been made over the subsequent 28-year period in understanding stereotypies sufficiently to use them as an index of

Table 1.1. Numbers of experimental papers[a] (classified by species and discipline) and reviews[b] on farm animal stereotypies: 1965–1992

	Numbers of papers	Total
Species		
Pigs	27	
Poultry	9	
Sheep	2	
Cattle	4	
Horses	5	
Discipline		47
Behavioural		
Descriptive[c]	15	
Experimental[d]	10	
Physiological	9	
Neural		
(pharmacological)	11	
Performance	2	
Reviews	16	16

[a] Only experimental papers in scientific journals have been included.
[b] Reviews with a major emphasis on farm animals.
[c] Descriptive behavioural studies (i.e. lacking a focused experimental design).
[d] Behavioural studies with a focused experimental design.

'welfare' and 'suffering'. As one crude guide to progress we have conducted a literature search using DIALOG (Dialog Europe, Oxford, UK) to access the BIOSIS data-base on papers relating to the study of stereotypies in farm animals between 1965 and 1992 (Table 1.1). We cannot guarantee that this represents a complete record (not all papers are available to Dialog), but we do believe it to be a close approximation. Most importantly, the survey shows the low number of published papers. Our data suggest that papers in this area have been published at the rather astonishingly low rate of approximately 1.9 per year. The vast majority (58%) of the experimental papers have dealt with pigs with 19% on poultry and 23% on the other species. The majority of the papers have focused on behaviour (53%) and the majority of these have been descriptive (60%) with no imposed experimental treatment structure. The next biggest class are pharmacological studies (23%) followed by physiological (19%) and lastly performance-related studies (4%). There have been 16 reviews with a major emphasis on farm animal stereotypies. We therefore estimate the ratio of reviews to experimental papers at 1:2.9.

There may be a number of factors that explain the low output of work on farm animal stereotypies. First, we would expect published papers to be relatively low in the early years as interest in the subject developed. In fact, output remained low for many years. Our survey indicated that output increased from 0.3 papers per annum between 1965 and 1974, to 2.1 papers per annum between 1975 and 1984 and finally to 3.6 papers per annum between 1985 and 1992. Second, the low output is also certainly a reflection of the overall amount of money being spent on animal behaviour and welfare research. It is only in recent years that research funding in farm animals has expanded from traditional animal science to support studies of farm animal behaviour and welfare. Third, there may also be an additional factor relating to the applied nature of the area. There has been a tendency within the field to 'talk' rather than to 'do'. This is indicated by the unbalanced ratio of reviews to experimental papers, and may in part reflect real or imagined 'political pressure' to come to definite decisions regarding the relationship between stereotypies and suffering. Further, it may be that the active pursuit of a practical 'goal' has hindered rather than enhanced progress by diverting attention away from work aimed at understanding the fundamental bases of stereotypies. For example, the tendency to only study animals under conditions in which they are kept by man undoubtedly reflects our tendency to see stereotypies as abnormal responses to specific 'housing systems' rather than in terms of fundamental underlying processes. Yet there is evidence that a number of factors such as level of specific motivation (Chapter 3), ability to voluntarily interact with the environment (Chapter 4) and individual differences in response to the environment (Chapters 2, 3, 6 and 7) are likely to influence development of stereotypies across a range of housing types. We would suggest, in fact, that stereotypies can only be effectively controlled by first understanding and then manipulating these underlying processes and not by *ad hoc* and unsystematic changes to the animal's environment.

Given this background, we have several justifications for a book on stereotypies at this time. First, we believe this is an opportune time to review the recent increase in published work (see above), not only to provide a synthesis of the findings but also to help direct future research effort on stereotypies (e.g. Chapter 8). Second, we were conscious that insufficient attention had been paid in the past to the heterogeneity of the different forms of stereotypy (Chapter 2). Third, we wished to present the study of stereotypies from the viewpoint of different disciplines. Too often the causes and function of stereotypies have been discussed by applied ethologists without recognition of the wealth of information available in other disciplines such as neurobiology. Behavioural studies alone cannot provide the same depth of understanding as when combined with physiology (Ladewig and von Borrell, 1988). Applied ethologists have been criticized for not appreciating the complexity of the neurochemical mechanisms underlying behaviour when hypothesizing about the causes of stereotyped behaviour (Dantzer, 1991). There is a substantial literature on stereotypies in species other than farm animals that can provide additional and perhaps supplementary data on stereotypies in farm animals. In this volume stereotypies are analysed in terms of motivation (Chapter 3), subjectivity (Chapter 4), physiology (Chapter 5), psychopharmacology (Chapter 6) and function (Chapter 7). Last we wish to suggest that practical decisions on whether and how to prevent stereotypies can only be made on the bases of an appropriate range of fundamental and applied research.

References

Command Paper 2836 (1965) *Report of the Technical Committee to Enquire into the Welfare of Animals Kept under Intensive Husbandry Systems*. HMSO, London.

Dantzer, R. (1991) Stress, stereotypies and welfare. *Behavioural Processes* 25, 95–102.

Harrison, R. (1964) *Animal Machines: The New Factory Farming Industry*. Vincent Stuart, London.

Hediger, H. (1955) *Studies of the Psychology and Behaviour of Captive Animals in Zoos and Circuses*. Butterworth, London.

Ladewig, J. and von Borrell, E. (1988) Ethological methods alone are not sufficient to measure the impact of environment on animal health and animal well-being. In: Unshelm, J., van Putten, G., Zeeb, K. and Ekesbo, I. (eds), *Proceedings of the International Congress on Applied Ethology in Farm Animals, Skara 1988*. Kuratorium für Technik und Bauwesen in der Landwirtschaft, Darmstadt, Germany, pp. 95–102.

Forms of Stereotypic Behaviour

GEORGIA J. MASON
University of Cambridge, Cambridge, UK.

Editors' Introductory Notes:
The division of the flow of behaviour into discrete (measurable) units remains a major methodological problem in animal behaviour. Mason starts by asking how we recognize stereotypies, pointing out that the defining characteristics of stereotypies do not provide a foolproof means of identification, and the division between stereotypies and other behaviour and between different types of stereotypies is not always clear.

Describing a behaviour as stereotypic implies that it shares common properties with other stereotypies. This assumption is most evident where comparisons are drawn between environmentally induced stereotypies and stereotypies arising through scheduled feeding or use of drugs. While not denying the importance of seeking general principles that operate across different circumstances, it is important to examine the degree to which different forms of stereotypy are analogous. In fact, as Mason points out, the behaviours that we describe as stereotypies are very diverse. Even within a 'class' such as environmentally induced stereotypies there is considerable heterogeneity. Mason discusses sources of this heterogeneity and comments on the relationship between the different classes of stereotypy, such as those induced by housing and those resulting from scheduled feeding. This chapter raises the question of whether the continued use of the category 'stereotypy' helps or hinders research.

Finally, Mason returns to the question of methodology, discussing the use of different recording methods and analysis techniques for stereotypies. She stresses the importance of presenting a complete description of the morphology, timing and stage of the stereotypy, as well as of the subject's circumstances, age and past history. More complete information, she argues, will allow a better understanding of the genuine equivalence of different stereotypies and consequently aid in the development of general theories and classifactory systems.

Introduction

Stereotypies are generally defined as unvarying, repetitive behaviour patterns that have no obvious goal or function (e.g. Fox, 1965; Hutt and Hutt, 1965; Ödberg, 1978). Variants of this definition are sometimes used; for example van Putten (1982) stipulates that the movements involved must be so abbreviated that they do not resemble any normal behaviour pattern, and Berkson (1967) allows stereotypies to be non-repetitive and hence includes behaviour such as abnormal posturing. However, the version first used by Fox, Hutt and Hutt, and Ödberg is now the consensual definition in the ethological and animal welfare literature (e.g. Wiepkema *et al.*, 1983; Dantzer, 1986; Fraser and Broom, 1990, p. 307; Mason, 1991a). This is not to say that stereotypies are easy to identify in practice. Two problems emerge when applying the definition. First, the borders of the category are blurred; how invariant, repetitive, or inappropriate must behaviour be before we call it stereotypic? Second, the broad definition covers such a multitude of behaviour types that to label something a 'stereotypy' is, in a sense, to say very little about it. In this chapter, I discuss the problems of identifying stereotypies, the reasons for their heterogeneity, and the implications this has for research.

Identifying Stereotypies: Where to Draw the Line

Distinguishing between stereotypies and other forms of behaviour can be problematic for three reasons. One is that there is often a continuum between stereotypies and the normal behaviour patterns from which they develop; deciding when behaviour has crossed the borderline is essentially arbitrary. Second, long-established stereotypies may become so 'ingrained' and habitual that an animal may incorporate them into otherwise normal behaviour patterns. Third, many behaviour patterns are stereotyped (i.e. invariant and repetitive) yet only a subset are considered stereotypies; ascribing 'abnormality' or lack of function to a behaviour pattern is often a subjective process (e.g. Dantzer, 1991).

THE DEVELOPMENT OF UNVARYING BEHAVIOUR

For the most part, stereotypies do not appear *de novo*; they develop. Therefore a continuum exists between proto-stereotypies and more developed, less ambiguous forms of the behaviour. The developing inflexibility of stereotypies is thought in some cases to arise with repetition (e.g. Meyer-Holzapfel, 1968; Fentress, 1976; Cronin et al., 1984). The form of the repeated behaviour pattern becomes less and less dependent on feedback from environmental factors (e.g. Lashley, 1921; Miller et al., 1960, pp. 81–93), and as these lose their role in modulating the movements, so variability decreases. External stimuli may still trigger the behaviour, but the exact form of the stereotypy is now 'self-organized' and independent of sensory feedback (Fentress, 1977). In other cases, the development of stereotypies may reflect other processes, such as learning, or increased dysfunctioning of the central nervous system (reviewed by Mason and Turner, 1993).

The continua above reflect long-term changes. However, a similar continuum between stereotyped and non-stereotyped versions of a behaviour pattern may also be seen in an individual over a much shorter timescale; in one animal, a behavioural pattern may look like a stereotypy in some circumstances but be far more variable at other times. This depends on the current state of the animal rather than on the behaviour's stage of development, and is a product of the animal being limited in the information it can take in from the environment and from its own limbs (e.g. Berlyne, 1960, p. 45; Fentress, 1976), and in the speed with which this information can be transmitted and processed (Carpenter, 1989, p. 335). This means that, in some circumstances, information from the periphery will make little impression on the performance of a behaviour pattern, so resulting in decreased variability. For example, if an animal directs its attention elsewhere while performing a behaviour pattern, less information from external factors relevant to that behaviour will be processed, and the pattern will become less flexible. This is seen, for example, in the way that mice groom their faces when in an unfamiliar environment: their behaviour becomes more stereotyped and is not affected by lesions that block sensory input from the facial area (Fentress, 1976). Fentress argues that this is because they are occupied in taking in information from their new surroundings. Similarly, if behaviour is performed at speed, movements will become too fast to be guided by sensory feedback (Lashley, 1917, 1951; Miller et al., 1960; Fentress, 1976; Carpenter, 1989, p. 235), and this too will reduce behavioural flexibility. For example, a caged hunting dog that performed a pacing stereotypy was more likely to trip over an obstacle in its path when it increased its speed (Fentress, 1976).

No rigid line therefore demarcates stereotypies. As a consequence, most authors do allow a certain degree of variation in the behaviour patterns they put into this category. For example, the body-rocking movements of mentally handicapped children change in amplitude, yet the movements are otherwise so similar, and each bout resembles every other so closely, that Berkson (1983) felt justified in calling the behaviour a stereotypy. However, the classification of other behaviour patterns is not so straightforward. For example, the abnormal behaviour of farmed mink can be rather variable. An animal might pace the length of its cage, sometimes rearing at one end, sometimes at the other, and occasionally visiting its nestbox. De Jonge et al. (1986) did not classify such patterns as stereotypies, yet I did, on the grounds that the movements involved were very predictable from bout to bout (Mason, 1992).

INCORPORATION INTO NORMAL BEHAVIOUR

A further stage in the development of some stereotypies is that they become elicited by a broad range of events, and are no longer dependent on the original causal stimulus alone (e.g. Fentress, 1976). Indeed, a stereotypy may become so readily performed that it is sometimes incorporated into the normal behaviour of an animal. In this way they are perhaps analogous to habits (Mason and Turner, 1993). For example, the head-swing developed by one stereotypically pacing dingo came to be shown whenever the animal made any turning movement, whatever the circumstances (Fox, 1971). In voles, too, a turning movement, once acquired, could be elicited by practically any sort of environmental variable (Hinde, 1970, p. 558, citing Fentress, 1965).

This raises problems when it comes to research. Should every incidence of such a movement be recorded as a stereotypy? Or alternatively, should the behaviour be scored as a stereotypy only when performed in particular circumstances, or in a repetitive bout?

WHEN IS A BEHAVIOUR PATTERN 'INAPPROPRIATE'?

In form and context, stereotypies appear to have no function or end-point. They are thus 'inappropriate' (Robbins and Sahakian, 1981), and in one sense of the word they are also abnormal (Fraser 1968; McMahon and McMahon, 1983, p. 37). Distinguishing a behaviour pattern with a function from one without is problematic. When a stereotypy is an exaggerated form of normal behaviour (Kiley-Worthington, 1977), at what point does it stop being adaptive? Body-rocking and thumb-sucking, for example, are shown by normal infant primates, including humans (Berkson, 1967, 1983; Thelen, 1979). The repetition is thought to aid normal motor and cognitive development (a form of behavioural 'scaffolding' (Bateson, 1986) that allows later adult behaviour to develop properly). In handicapped children, the body-rocking they display may represent the retention of these normal movements. The movements may remain because alternative, more adult, competing modes of functioning are unable to develop, and they may continue to have a function (Berkson, 1983). So should such movements be called stereotypies? Similar problems are discussed by Robbins and Sahakian (1981), Fraser and Broom (1990, p. 311) and Rushen et al. (1990); for a recent review see Mason (1991a).

This is more than just a practical difficulty; it points to a fundamental problem in the basic definition of stereotypies. Two of the criteria, rigidity and repetitiveness, are widespread characteristics, common to grooming, ingestive behaviour, displays and learnt motor skills (e.g. Lashley, 1921;

Miller et al., 1960, pp. 81–93; Hinde, 1970, pp. 19–22; Martiniuk, 1976, pp. 27–28 and 142–144; Dawkins, 1986), behaviour patterns whose causes, goals and functions are all very different. To use rigidity and repetitiveness to link behaviour patterns is therefore akin to using an ancestral characteristic to group species together (see e.g. Ridley, 1986, pp. 56 and 61); it could lead to the behavioural equivalent of uniting snakes and slow-worms in one group and birds and lizards in another, on the grounds of pentadactyly. So is there anything that uniquely defines stereotypies? Their other defining characteristic is their lack of apparent function, but this is clearly difficult to judge (cf. Chapter 7), and, furthermore, is it biologically meaningful to group things on the basis of what we do not know about them? Thus the basic definition of stereotypy does not require or imply that the behaviour patterns meeting the criteria are homologous. Indeed, considerable evidence suggests that they are not.

The Heterogeneity of Stereotypies

Stereotypies differ greatly from each other. Some are induced by the environment, and others by pharmacological or surgical interference with the central nervous system, while others, in humans, are associated with psychiatric conditions, the organic bases of which may remain obscure. They are heterogeneous in their physical appearance and in when they are performed, and they seem to arise from a range of source behaviour patterns. They are also heterogeneous in the manner in which they meet the defining criteria; stereotypies may be invariable and repetitive to very different extents, and these two features may vary independently of one another. In the sections that follow, I discuss how stereotypies differ in cause and in effect, and other sources of individual variation in the behaviour.

STEREOTYPIES INDUCED BY THE ENVIRONMENT

The types of environments that give rise to stereotypies are considered in Chapter 3. These environments suggest that stereotypies are often (though not always – see Mason, 1991b) an indicator of poor welfare. A variety of stereotypies is elicited by the environment. For example, they range in form from the energetic and acrobatic, such as somersaulting in chipmunks (pers. obs.) and jumping in voles (Fentress, 1977; Ödberg, 1986), to movements as restrained and minimal as eye-rolling in veal calves (Fraser and Broom, 1990, p. 312). Some stereotypies are idiosyncratic, while others are species typical. For example, most mink that perform stereotypies pace, but a few perform somersaults (Mason, 1992). Stereotypies may be conspicuous and unmistakable, or more cryptic and complex, with a long periodicity (e.g. Rushen, 1984).

In their natures, stereotypies range from the transient (e.g. a human pacing in a waiting-room (Ridley and Baker, 1982) or a newly caught wild fox pacing in its trap (White *et al.*, 1991)), to the long-term (e.g. the stereotypies of blind children (Stone, 1964)). Bouts may also be easy or difficult to interrupt. For example, some polar bears seem to remain alert while pacing and can be interrupted even by a visitor dropping a leaf into their moat (Wechsler, 1991), while others continue to pace even if food is placed in the enclosure (Meyer-Holzapfel, 1968, citing Eipper, 1928). Stereotypies may be unlearnt species-specific behaviour patterns, such as that developed from the courtship ritual of a male okapi (Ödberg, 1978), or they may be conditioned responses, such as the stereotyped sequences that develop in key-pecking pigeons rewarded with food (Schwartz, 1980, 1982). Some appear harmless, e.g. the digit-sucking seen in human babies *in utero* (Sroufe and Cooper, 1988, p. 114), while others, such as self-biting (Berkson, 1968; Broom, 1983), are injurious.

Stereotypies have been classified into a number of subgroups, according to the circumstances of their development and their current characteristics. The various schemes that have been proposed to categorize stereotypies have been discussed and criticized by Mason (1991a). Rather than emphasizing again the extent of the differences between stereotypies, and perhaps leaving the reader with a feeling of dismay, I would like instead to discuss the reasons for this great diversity of form. Identifying the sources of this diversity may help to determine the research methods used to study stereotypies, and help to develop further classificatory schemes for this behaviour.

REASONS FOR HETEROGENEITY

The age of the subject

The form and persistence of a stereotypy is often determined by the age of the subject when exposed to the eliciting environment. For example, the effects of caging seem more profound on a young than on a mature animal (e.g. Berkson, 1968). The stereotypies that develop in young animals (Berkson, 1967; Ridley and Baker, 1982) and institutionalized children (e.g. Mason and Green, 1962) are known as deprivation stereotypies, and they are in general very difficult to abolish (Ridley and Baker, 1982), sometimes persisting even when the subject is put into a normal environment (e.g. Davenport and Menzel, 1963; Meyer-Holzapfel, 1968). They tend to be non-locomotory, and may involve self-mutilation (Ridley and Baker, 1982). They are also exacerbated by stimulant drugs, at least at low doses (e.g. Berkson and Mason, 1964; Lyon and Robbins, 1975, citing Fitz-Gerald, 1967; Robbins and Sahakian, 1981; Benus *et al.*, 1991). In contrast, animals caged or humans institutionalized as adults develop stereotypies that usually can be altered or eliminated by changes in the environment (Draper and Bernstein, 1963;

Berkson, 1967; Ridley and Baker, 1982). These stereotypies are not increased (Ödberg, 1984a), and indeed are sometimes abolished (Ridley and Baker, 1982), by stimulant drugs.

The stereotypies developed by young animals may be more severe because their central nervous systems are more plastic, and hence are more profoundly altered by their environment. Those developed in adults, in contrast, may be the product of a normal central nervous system (Ridley and Baker, 1982). Young animals may also be especially sensitive because they are primed to develop stereotypies by aspects of the environment that affect them more greatly than adults. For example, isolation may predispose them to developing certain forms of stereotypy because in the absence of their mother, they may be more affected by things they perceive as frightening (Harlow and Zimmerman, 1959; Bowlby, 1988, p. 27, pp. 121–122; Martin and Bateson, 1988; Suomi, 1989).

The form of the stereotypies developed will also be influenced by age, for two reasons. First, an animal's stage of physical development may be important; Berkson (1968) suggests that body-rocking does not appear in very young primates because it requires a certain degree of motor competence. The complexity of stereotypies may then increase with age as the result of increased maturity and muscular coordination (Bernstein and Mason, 1962). Second, the animal's current behavioural repertoire, which will be influenced by its age, will also affect the forms of stereotypy that arise. A young animal may develop a sucking stereotypy, for example (e.g. Kiley-Worthington, 1977), but it is unlikely to develop a scent-marking stereotypy, as seen in one zoo-housed pine marten (Hediger, 1950, p. 75). Age-related differences in stereotypy can be seen in the cow. Early weaved veal calves develop a form of stereotyped tongue-rolling that involves the end of the tongue curving back in to the mouth and being sucked on, before being again extended. This seems to reflect a suckling deficit, and contrasts with the behaviour of adult cows whose tongue stereotypies include components of the movements involved in pulling up vegetation (Fraser and Broom, 1990, p. 314).

Stereotypies develop not only from normal behaviour, but from behaviour patterns that develop as if to substitute for absent sources of stimulation. Again, these will be influenced by the individual's stage of development. For example, Mason and Berkson (1975) present evidence that the body-rocking developed by isolation-reared primates develops as a substitute for the proprioceptive and kinaesthetic stimulation that they would normally receive on being carried by their mothers, stimulation that may be necessary for potentiating the development of more mature behaviour, such as play. Providing young rhesus monkeys with an artificial mother that moved abolished the behaviour. Similarly, congenitally blind or mentally disabled humans develop stereotypies that are not seen in adults suddenly stricken with the same handicap. Ridley and Baker (1982) suggest that these stereotypies arise from the need for sensory input during maturation of the central nervous system.

Experience

Past experience may affect the levels of stereotypy developed. If the animal has been reared in a barren or otherwise impoverished environment, it may be more generally prone to inflexible behaviour. For example, isolation-reared rats are unusually persistent in learned tasks, compared with socially reared animals (Morgan, 1973). A general inflexibility in behaviour induced by rearing conditions could predispose it to the later development of stereotypies. More specifically, the early loss of an attachment figure, such as the mother, may make individuals more psychologically sensitive to later adversity (Rutter, 1981, pp. 180 and 192, and Bowlby, 1988, p. 36, reviewing the work of, for example, Brown and co-workers), and hence predispose them to stereotypies in later life.

Social experience may also be influential, if it involves conspecifics that themselves perform stereotypies. The rate of stereotypy development may be accelerated if an animal's neighbours show the behaviour (Palya and Zacny, 1980). For example, weaving in horses tends to spread throughout a stable (Kiley-Worthington, 1983).

The past experience of an animal may also shape the forms of stereotypy developed. The patterns that arise may stem from behaviour patterns that have been learnt previously. For example, autistic children may play the same record again and again, in a stereotyped manner (Marks, 1987, p. 420), and sows develop stereotypies from the learnt behaviour pattern of drinker-pressing (e.g. Terlouw *et al.*, 1991a). Experience may also affect the behaviour patterns likely to give rise to stereotypies by influencing what an animal finds particularly motivating. For example, tame animals are probably less likely than captive wild animals to develop stereotypies arising from attempts to escape the presence of humans (Meyer-Holzapfel, 1968). Cronin (1985) suggests that in young sows with little or no prior experience of tethering, this form of restraint leads to frustration and redirected aggressive behaviour, from which develop stereotypies such as bar-biting. The older, more experienced sows tend to react instead with self-directed activities such as oral stereotypies, which may represent some kind of self-stimulation. (Not all researchers agree with this interpretation, however; see Chapter 3.)

The eliciting circumstances

The current environment affects the stereotypies that are developed in three ways. It will elicit some behaviour patterns more than others; it will affect the level or intensity with which these behaviour patterns are performed; and, finally, it will affect the precise form that these behaviour patterns can take. For example, the exact morphology might be constrained by how much room there is in the enclosure (Levy, 1944).

If the environment is generally unstimulating, stereotypies may arise from attempts to increase sensory input. For example, a study of two mentally handicapped children showed that they spent much time performing stereotypic hand-gazing. The children were then diagnosed as being very short-sighted, and when provided with glasses their behaviour showed a dramatic improvement. They held their heads upright, looked around them at what was going on in their environment, and reduced their stereotypies considerably (Gallagher and Berkson, 1986). It was as if their stereotypies had been one of the few sources of stimulation in their previous fog of myopia.

Cage-induced stereotypies can often be traced to specific motivations (see Chapter 3). For example, the stereotypic biting patterns seen in monkeys prior to a regular electric shock (Hutchinson, 1977) derive from defensive aggression (see Bolles, 1970), while Rushen (1984, 1985) suggests that in sows, pre-feeding stereotypies such as bar-biting represent frustration. Escape attempts give rise to the stereotypies of voles in a laboratory cage (Ödberg, 1986) and also the pacing of laying hens against the door of their cage, when frustrated by the absence of nesting sites (Duncan, 1970; Wood-Gush, 1972). Nest-building behaviour can also be the source of specific stereotypies. Finches deprived of nesting material develop stereotypies involving the plucking or carrying of their own feathers (Hinde, 1958). Feeding and foraging behaviour is also often a source of stereotypy. For example, the repetitive vomiting and re-ingestion of food by a sloth bear, when chased from her food by an aggressive cagemate (Meyer-Holzapfel, 1968), may have substituted for the consummatory behaviour she was unable to perform normally. Appetitive behaviour may give rise to stereotypies if *ad lib* food is provided with little need to forage (Keiper, 1969), or if the animal has a regular period of expectancy before each meal. For example, experimentally induced stereotypies develop in many species during a regime of periodic feeding, where food deliveries are evenly spaced and, as the time of food delivery approaches. They are known as 'terminal' stereotypies, for they are seen at the end of the inter-food interval (Staddon and Simmelhag, 1971; Anderson and Shettleworth, 1977), are very species-typical (e.g. spot-pecking in pigeons), and individuals differ little. The stereotypies seem almost automatic, arising quite quickly and occurring even if they reduce the rate of food delivery by interfering with an operant response (Staddon and Simmelhag, 1971). They may well be analogous to the pre-feeding stereotypies seen in some farmed and zoo-housed animals, although the equivalence may depend on the species.

However, environment-induced stereotypies are not always clear in their motivational origins. For example, the pacing of polar bears in zoos does not seem fully explicable in terms of appetitive or territorial behaviour (Wechsler, 1991). In addition, the motivation behind the 'wall-writing' stereotypy of one female chaffinch kept without nesting material (Hinde, 1958) is obscure; why did she not develop a more obvious alternative to nest-building, like the birds described above? Finally, a second form of experimentally induced stereotypy is seen on a regime of periodic feeding, additional to the 'terminal' type and much less easy to explain. These idiosyncratic stereotypies appear on the regime in the middle of the period between food deliveries, and are known as 'interim' activities.

Interim stereotypies are heterogeneous, and their origins are not entirely clear. They include polydipsia (excessive drinking), behaviour that may represent escape attempts (Staddon and Simmelhag, 1971), and a number of other patterns that may stem from displacement activities (Staddon and Simmelhag, 1971; Rushen, 1984) due to the disinhibition of low priority behaviour patterns when higher priority ones cannot be performed (Anderson and Shettleworth, 1977; McFarland, 1985, p. 385). The heterogeneous nature of these behaviour patterns, and the unlikelihood of a unitary explanation even for schedule-induced polydipsia (SIP) alone, has been emphasized by Roper (1983; see also Anderson and Shettleworth, 1977). In the stereotypy literature, links have sometimes been made between cage-induced stereotypies and schedule-induced behaviour such as SIP (e.g. Dantzer and Mormède, 1981, 1983; Rushen, 1984). However, some caution may be required; for example, schedule-induced stereotypies are not seen if the interval between reinforcements is greater than a few minutes (e.g. Falk, 1971; Roper, 1980), and on these grounds appear unlikely to be analogous with anything seen in farms or zoos. Indeed, experiments to investigate the links between spontaneous stereotyped chain-chewing in sows and schedule-induction of that behaviour suggest basic differences between these two classes of behaviour (Nielsen et al., 1993).

Once developed, various environmental factors can influence the morphology of a stereotypy. In particular, it can vary in form with the strength of the eliciting stimuli. It might become simplified, if the eliciting stimuli are less intense. For example, the somersaulting of one rhesus monkey was reduced to mere head movements in a larger cage (Draper and Bernstein, 1963). Alternatively, it might change qualitatively when arousal increases, perhaps reverting to an older version of the behaviour. For example, a caged hunting dog, when alarmed, would revert to an old form of stereotypy, leaping to clear a chain no longer there (Fentress, 1976). Current environment also influences the level of the behaviour. Appleby et al. (1989) showed that, in sows, the performance of stereotypy was encouraged by the noise of neighbours chain-chewing.

Species

A subject's species obviously exerts an enormous influence on the sort of behaviour that develops into a stereotypy: for example, hunting animals pace, while grazers roll their tongues, and so on. This is as true of deprivation-induced stereotypies as it is of cage-induced ones. Among isolation-reared primates, rocking is almost always accompanied by self-clasping in monkeys but this behaviour seems to be 'optional' in humans (Berkson, 1967), a difference that presumably reflects the role that clasping the mother plays in the early life of non-human primates.

Some species differences are not so easy to explain in terms of normal behavioural repertoire, however. For example, in a dark, barren container, two species of vole develop different typical movements. *Microtus agrestis* tends to weave and circle around its drinking spout, *Clethrionomys britannicus* tends to jump vertically (Fentress, 1977). Also, the number of different stereotypies seen in isolated primates is greater in apes than in monkeys, a phenomenon that Berkson (1968) attributed to 'phylogenetic development', but could not explain more fully. In schedule-induced stereotypies, too, species differences are difficult to explain. For example, SIP is seen in rats but not pigeons or gerbils, but rats do not display the schedule-induced aggression seen in pigeons nor the wheel-running evident in gerbils (reviewed by Roper, 1983).

The age of the stereotypy: changes in nature

The age, experience and environmental circumstances of an individual do not determine all the characteristics of its stereotypies. For example, highly persistent stereotypies, little affected by environmental change, can develop in adult animals as well as in the young (e.g. Hinde, 1958). The explanation may lie in how stereotypies change when performed over a long period of time. As a stereotypy becomes established in the behavioural repertoire of an animal, its nature can change in a number of ways.

As discussed earlier, established stereotypies may be elicited by a broad range of events (Fentress, 1976; Ödberg, 1978), and hence performed outside the original eliciting situation. The uncoupling of performance from the original eliciting factors is known as emancipation (e.g. Groothuis, 1989). Such stereotypies tend to be performed at times when there is little competition from other, higher priority, behavioural systems (e.g. Hinde, 1962, 1970; Duncan and Wood-Gush, 1974, interpreting the data of Keiper, 1970). They are also performed when general arousal increases (Fentress, 1973, 1976; Ridley and Baker, 1982). For example, the established stereotypies of squirrel monkeys can be elicited through electrical stimulation of the mesencephalic reticular system (Fentress, 1976).

Established stereotypies may even be elicited or facilitated by stimuli that would have interrupted them earlier in their development (Fentress, 1977). Indeed, they are much more difficult than developing stereotypies to discourage or interrupt by changing the environment (Kiley-Worthington, 1977; Cronin *et al.*, 1984). For example, once firmly established, the pacing shown by food-deprived hens faced with a covered food-dish does not disappear even when food is made available (Duncan and Wood-Gush, 1972).

Established stereotypies are also no longer inhibited by tranquillizers (Feldman and Green, 1967; Duncan and Wood-Gush, 1974). This could suggest that their elicitation does not depend on emotional, limbic systems (e.g. Kennes and Ödberg, 1987; Ödberg, 1989) but, rather, that they are emotionally neutral (e.g. Duncan and Wood-Gush, 1974), or 'pure motor automatisms' (reported in Kennes and Ödberg, 1987; Ödberg, 1989; also Dantzer, 1986). Nor are they inhibited by opioid receptor blockers, but seem to be essentially under dopaminergic control (reported in Cronin et al., 1985a; Kennes et al., 1988; see also Chapters 6 and 7).

It is unclear whether all these developmental changes in a stereotypy occur synchronously (Kennes et al., 1988). The change in emotional involvement might precede emancipation; for example, Cronin et al. (1984, 1985b) described how the post-feeding stereotypies of tethered sows change from short bursts of apparently aggressive activity to prolonged bouts of much less aggressive behaviour. Only if restraint continued did this behaviour then come to be performed at additional times of day.

The age of the stereotypy: changes in form

Stereotypies may also change in form with age. They may lose elements of the original behaviour pattern. For example, a tic-like head-movement developed from the abbreviated intention movements of nest-material collection, in a caged chaffinch (Hinde, 1962). Such stereotypies are sometimes known as 'incomplete' (van Putten, 1982; Broom, 1983). The abbreviation may be due to cumulative repetition and the rapidity with which they are performed (Dantzer, 1986). The loss of elements might be responsible for the reported transition from environment-directed stereotypies such as bar-biting, to self-directed stereotypies such as sham-chewing, in tethered sows (Dantzer, 1986). A similar change has been reported in the horse, with crib-chewing sometimes progressing to air-sucking (Fraser, 1980, p. 246), and weaving changing from a side-to-side pacing to a stationary head-swing (Meyer-Holzapfel, 1968). However, stereotypies do not always simplify or become less environment-directed with age. For example, head-banging may develop in human infants who previously had shown head-rolling or body-swaying (Kravitz et al., 1960). The degree to which behaviour is abbreviated, if at all, may depend to some extent on the length or complexity of the movements involved (e.g. Hudson, 1969).

While the movements involved may become abbreviated, the sequences of movements may become more complex with age. Sow stereotypic sequences grow longer with continued restraint, as new components are added (Cronin et al., 1984). Similarly, the complexity of mink stereotypies increases with age (de Jonge et al., 1986).

STEREOTYPIES INDUCED BY INTERFERENCE WITH OR DAMAGE TO THE
CENTRAL NERVOUS SYSTEM

Stimulant drugs

Stereotypies are induced by psychomotor stimulant drugs such as amphetamine and apomorphine (Lyon and Robbins, 1975; Robbins and Sahakian, 1981; see Chapters 6 and 7). The animal repeats movements in an unvarying repetitive way, and the bouts of such stereotypies are similar in appearance each time they are performed; an individual's characteristic stereotypy may be shown on successive amphetamine trials even if these are months apart (Lyon and Robbins, 1975). Locomotor activity also increases on treatment with a stimulant (Robbins and Sahakian, 1981). In human subjects under the influence of amphetamine, routine and thought processes become unvarying and obsessive or compulsive, in a manner similar to that of schizophrenics (Robbins, 1976; Robbins and Sahakian, 1981; Ridley and Baker, 1982).

Stimulant-induced stereotypies share various similarities with environment-induced forms of the behaviour. For one thing, they seem very compulsive and will persist even when punished (Lyon and Robbins, 1975, reported by Robbins and Sahakian 1981). They may also involve restricted attention (Robbins and Sahakian, 1981; Robbins *et al.*, 1990) and a general tendency to persist with all behaviour patterns (Ridley and Baker, 1982). In addition, stimulant-induced stereotypies are often potentiated by arousal. For example, rats show a circadian rhythm in apomorphine stereotypies, with enhanced responses to the drug at certain times of day (Robbins and Sahakian, 1981, citing Nakano *et al.*, 1980) and, likewise, caged voles have a circadian rhythm in their environment-induced stereotypies, peaking when the animals are most active (Ödberg, 1986). Pinching a rat's tail, and depriving it of food, both enhance stimulant stereotypies (Antelman *et al.*, 1979, reported by Robbins and Sahakian, 1981). Indeed, the stimulants may be acting as stressors themselves (e.g. Mittleman *et al.*, 1986). Admittedly, in some cases stressful treatments do not enhance the development of stimulant stereotypies, and may even inhibit or disrupt them (Sahakian and Robbins, 1975; Robbins and Sahakian, 1981 citing e.g. Lyon and Randrup, 1972; Cabib *et al.*, 1985). However, this may be similar to the way in which very intense stimuli can interfere with environment-induced stereotypies (Fentress, 1977; Robbins and Sahakian, 1981). Finally, dopaminergic systems are involved in the behaviour, and opioids are involved in some cases (Robbins and Sahakian, 1981), as they are in some environment-induced stereotypies (Ödberg, 1984a; Cronin *et al.*, 1985a; Kennes *et al.*, 1988). The neurological bases of these stereotypies are discussed in more detail in Chapters 6 and 7.

It is also worth repeating here that, in some cases, stimulant drugs may increase the performance of an animal's deprivation stereotypies. However, other forms of environment-induced stereotypy, developed in adults, are not enhanced by the administration of stimulants (Berkson and Mason, 1964; Lyon and Robbins, 1975; citing FitzGerald, 1967: Ridley and Baker, 1982; Ödberg, 1984a).

Classifying the behaviour patterns induced by drugs can be as problematic as identifying environment-induced stereotypies. Drug-induced behaviour patterns on the borderline of stereotypy are of two types. In one, the hyperactive animal moves along a very fixed path, yet the movements involved in the locomotion are not themselves stereotyped (Robbins and Sahakian, 1981). In the other, some of the movements induced are highly predictable, yet the animal shows considerable flexibility of overall behaviour in order to perform them. For example, an amphetamine-treated guinea-pig would always gnaw on one spot on the coat of a cagemate. To do this, it would often have to change posture and position (Sahakian and Robbins, 1975, discussed in Robbins and Sahakian, 1981). Whether such behaviour patterns should be called stereotypies is a matter of judgement.

In addition, like stereotypies induced by the environment, stimulant-stereotypies are a heterogeneous group. This is graphically demonstrated by their complex relationships with environment-induced stereotypies. Individual differences between sows in their behavioural response to amphetamine were related to development of stereotypic chain-chewing under restrictive feeding and housing conditions (Terlouw et al., 1993a). Drug-induced oral stereotypies were negatively correlated with the chain-manipulation, while the drug-induced locomotory activity positively correlated with the environment-induced stereotypy. The varying properties of different drug-induced stereotypies have also been demonstrated in other experiments, where they have been related to the precise nature of the causal agent, the nature of the animal's environment and the characteristics of the individual animal.

The precise nature of the stimulant affects the typical appearance of the resulting stereotypy. For example, rats given amphetamine tend to react initially with an increase in locomotion, but this does not occur with apomorphine (reviewed by Robbins and Sahakian, 1981). Apomorphine is also characterized by the prevalence of oral movements in the resultant stereotypies (Robbins et al., 1990). However, species differences are also important here as many of these differences between amphetamine and apomorphine reported in rats are reversed in pigs (Terlouw et al., 1993b) The stereotypies induced by different drugs also differ in their response to a dose increase (Fray et al., 1980), and the way in which they are affected by, for example, food-deprivation (see e.g. MacLennan and Maier, 1983). They may also involve opioid systems to different extents (see Robbins and Sahakian, 1981). Ödberg (1978) reviews evidence that lesions of the corpus striatum inhibit amphetamine stereotypies but not apomorphine ones.

Like environment-induced stereotypies, stimulant-induced stereotypies also vary according to the magnitude of the causal stimulus. Stereotypies become more intense with increasing doses of stimulant, their rate of repetition increasing, and the movements becoming increasingly abbreviated (e.g. Lyon and Robbins, 1975; Robbins and Sahakian, 1983). As the stereotypy intensifies, the movements become more and more dissociated from their original context, in timing and in orientation (Robbins, 1976; Robbins and Sahakian, 1981; Robbins, 1982). They also become species-typical instead of idiosyncratic (e.g. as reviewed by Lyon and Robbins, 1975), typically involving oro-facial or isolated limb movements (e.g. Robbins and Sahakian, 1981). At high doses the stereotypies may become self-damaging, involving, for example, self-biting and over-grooming (Ridley and Baker, 1982).

Local circumstances also affect the development of stimulant-induced stereotypies, and are therefore partly responsible for the different forms seen (e.g. Sahakian and Robbins, 1975; Robbins *et al.*, 1990). Drug-induced behaviour is influenced by both physical and social aspects of the environment. Group-housed dogs may develop the stereotypic following of other animals, when under the influence of amphetamine, and cats in a closed environment tend to develop sniffing stereotypies, those in an open environment, 'looking' ones (Ellinwood and Kilbey, 1975). Animals may be influenced by even more subtle factors, such as the site of administration of the drug. Rats injected with apomorphine in the flank tend to develop gnawing stereotypies, while those injected in the neck tend to develop locomotor movements (Robbins and Sahakian 1981, citing Ljunberg and Ungerstedt, 1977).

The role of the environment demonstrates how individual differences in prepotent behaviour affect the type of stereotypy developed. Stimulant-induced stereotypies tend to develop from behaviour that was predominant prior to treatment, and experience plays an important role here. Experience is therefore another source of differing stereotypies. Food-deprived animals, for example, are likely to develop eating stereotypies when treated with stimulants (Ellinwood and Kilbey, 1975; Robbins, 1982; Dantzer, 1986, reviewed by Robbins *et al.*, 1990). Past learning can also be a source of stereotypies. Some amphetamine-addicts develop complex stereotypies involving the dismantling and reassembly of mechanical equipment (Ridley and Baker, 1982), and rats trained to lever-press to avoid a shock develop a lever-touching stereotypy on amphetamine (Lyon and Robbins, 1975). Other aspects of experience are also important. For example, amphetamine induces higher levels of stereotypy in rats that have been isolation-reared (e.g. Sahakian *et al.*, 1975; Einon and Sahakian, 1979). Previous experience of the drug itself can also influence both the form of the stereotypy and the levels displayed.

The morphology of stereotypies can be affected by prior experience of the drug due to conditioning. This explains why individuals may develop characteristic stereotypies that recur from one trial to the next, and why amphetamine stereotypies can be induced in an 'experienced' animal by administration of a placebo (Lyon and Robbins, 1975). Prior experience of the drug also influences the form and the level of stereotypy developed in another way. Stereotypies become more intense with repetition of the treatment, as the animal becomes sensitized to the drug (e.g. Robbins and Sahakian, 1981). This change resembles the change produced by increasing doses, as discussed above. The stereotypies seen in the sensitized animal are less easy to interrupt with external stimuli (Robbins and Sahakian, 1981), and they seem to be more self-organized, with less reliance on environmental factors (Ellinwood and Kilbey, 1975; Robbins and Sahakian, 1981). The abbreviation seen perhaps parallels the simplification seen during the development of some environment-induced stereotypies (Dantzer, 1986). This transition is not a simple process. The various stereotypies elicited by increased or repeated dosage of one drug show different patterns of sensitization (Eichler et al., 1980), and may, as they change from complex to more simple forms, have different neurophysiological bases (reviewed by Robbins and Sahakian, 1981; Robbins and Sahakian, 1983; Mittleman et al., 1986; see also Chapters 6 and 7).

Finally, the age and species of the animal also play a role in determining the nature of its stimulant stereotypies. For example, rats typically sniff and make head and mouth movements when treated with amphetamine, while primates make hand movements (reviewed by Robbins and Sahakian, 1981) and pigeons peck (Lyon and Robbins, 1975). The effect of the subject's age has been demonstrated in the pig. Apomorphine leads to stereotyped locomotion with the snout in contact with the ground, along with stereotyped drinking and licking movements, in adult pigs (Terlouw et al., 1993b), but leads predominantly to snout-rubbing stereotypies in piglets (Fry et al., 1981).

Other drugs

Other drugs that intereferc with central dopamine systems of the brain can also result in stereotypies. L-dopa results in tics, stereotypies and compulsive behaviour akin to that of schizophrenia (e.g. graphically described in Sacks, 1990; also Dantzer, 1991). Morphine and other opioids also lead to stereotyped locomotion and oral stereotypies, when given to rats and mice (reviewed by Robbins and Sahakian, 1981).

Damage to the brain

Stereotypies can result from physical damage to the brain. For example, destruction of the frontal lobes disrupts planned, anticipatory, goal-directed behaviour and an individual's ability to monitor whether or not behaviour patterns have been appropriate (e.g. Luria, 1973, pp. 89–90). One manifestation of this disability is that the subject may break off from complex behaviour sequences before completion, and instead perform old responses, now inappropriate, or react impulsively to stimuli in the immediate environment (Stuss and Benson, 1984; Frith and Done, 1990; Robbins, 1991). This can result in transient stereotypies. For example, a dog with a bilateral frontal lobotomy given two bowls of food, may constantly, purposelessly, shift between the two (Luria, 1973, pp. 89–90).

Animals with bilateral hippocampal damage develop stereotypies in a rather different manner. They are characterized by generally 'excessive' and unvarying behaviour (e.g. Devenport et al., 1981). Stereotypies are the most extreme form of this, especially prior to the arrival of food, when the behaviour may be superstitious (Devenport et al., 1981, reviewed by Robbins et al., 1990). Hippocampal damage also exaggerates the head and neck stereotypies seen in amphetamine-treated rats (Robbins et al., 1990).

This hyperactive behaviour may be the result of unmodulated catecholamine activity. For example, the loss of the inhibitory input of the hippocampus to areas of the basal ganglia results in the type of hyperactivity that is also produced by amphetamine (Robbins and Everitt, 1982). A number of further behavioural changes may be responsible for the specific development of stereotypies. Hippocampectomy results in the reduced use of novel environmental cues (reviewed by Boucher, 1977). It also reduces levels of exploration (reviewed by Boucher, 1977) and hence reduces spontaneous alternation behaviour in a maze (Berlyne, 1960, pp. 128–143; Blozovski, 1986). Perhaps as a consequence of their reduced attention to novel external factors, animals that have had a hippocampectomy are less distractible than intact animals when thoroughly engaged in a task, showing reduced or even absent orientation reactions to presented stimuli (Pribram and McGuiness, 1975).

Examples from human clinical psychology

Studies by psychologists and psychiatrists reveal the association of stereotypies with a number of conditions in humans, particularly autism, schizophrenia and learning disabilities. For example, autistic children commonly engage in arm-flapping and object-twiddling, among other stereotypies (e.g. Lovaas et al., 1971; Frith and Done, 1990).

These human stereotypies often share features of animal stereotypies. Enriching the environment, for example, reduced the stereotypic rocking of individuals with learning disabilities (Forehand and Baumeister, 1971). Another similarity is that the stereotypies associated with some clinical conditions may become abbreviated with time; for example, those of schizophrenics are often greatly curtailed versions of a previous behaviour pattern (e.g. Ridley and Baker, 1982). One schizophrenic patient was reported to repeatedly show passers-by a piece of paper on which had once been written her address, and make sounds rooted in requests to take her home (Frith and Done, 1990).

Once again, these stereotypies are heterogeneous. For one thing, their morphologies vary widely. Schizophrenics differ from the autistics described above in that their stereotypies often involve characteristic facial movements

(Frith and Done, 1990), while individuals with learning disabilities often make simple body movements (e.g. Berkson, 1983). In some cases, environmental enrichment does not decrease the levels shown. For example, the stereotypies of autistic children may increase with environmental complexity (Hutt and Hutt, 1965). Nor are all stereotypies abbreviated with time. Progressive abbreviation is not seen in those of autistic children (Garrigues *et al.*, 1982). The correlates of this behaviour also differ greatly from one case to the next. Autistics are characterized by a desire for sameness, manifest in their tendencies to arrange objects in ordered patterns, to ask the same questions again and again, and to insist on daily routines (e.g. Prior and MacMillan, 1973; Baron-Cohen, 1989). This may encourage the performance of stereotypies in this particular group of subjects. The relationship between stereotypy levels and IQ also differs between different conditions. Stereotypy levels reduce with increasing IQ in those with learning disorders (e.g. Down's syndrome individuals; Frances, 1966), but this is not the case in autistic children, where the only change is an increase of the complexity of the movements performed (Frith and Done, 1990). A final source of differences in the stereotypic behaviour of humans is the range of cognitive correlates that are associated with it. Such correlates would, of course, be inaccessible in studies of animals. Tics, as are seen in Tourette's syndrome amongst other conditions, are quite involuntary (Frith and Done, 1990). Neurotic obsessive behaviour, such as repetitive hand-washing, is associated with a feeling of compulsion, but also an awareness that the behaviour is without real function (Rachman, 1985); and psychotic compulsions may have elaborate and imaginative superstitious rationales. For example, one schizophrenic patient believed that repeatedly touching his ear made his blood pump round (Jones, 1965).

Stereotypies that differ in their consequences

Another aspect of heterogeneity arises not from the physical forms of the behaviour, nor the circumstances in which they develop, but from the consequences of their performance. Stereotypies differ in their emergent properties – a fact of considerable welfare significance (e.g. Mason, 1991b).

Stereotypies that harm

Stereotypies may have costs. They may consume energy (e.g. Cronin *et al.*, 1986); for example, locomotory stereotypies can cause weight loss in horses (Fraser, 1980, p. 247; Fraser and Broom, 1990, p. 309) and mink (Mason, 1992). The repeated movements can also have more specific physical effects. For example, wind-sucking and crib-biting reduce the condition of stabled horses (Fraser, 1980, p. 246; Kiley-Worthington, 1983), wind-sucking causing gastrointestinal catarrh, colic and reduced food intake (Dodman *et al.*,

1987; Houpt, 1987; Fraser and Broom, 1990, p. 311), and crib-biting causing tooth-wear (Broom, 1983; Dodman *et al.*, 1987; Houpt, 1987) and the ingestion of splinters (Dellmeier, 1989). Repeated contact of the body with cage walls or bars during stereotypy may also cause sores (Morris, 1964; Meyer-Holzapfel, 1968; Ödberg, 1986), or impact injury (Fraser and Broom, 1990, p. 310). The stereotypies of autistic children are also detrimental, but psychologically, rather than physically, in that they may prevent learning and social interaction (Lovaas *et al.*, 1971; Koegel and Covert, 1972).

Stereotypies that benefit

Some stereotypies, conversely, seem to have benefits, or at least to be reinforcing. They may be adequate substitutes for the normal behaviour patterns from which they have arisen, or they may have more general roles in helping the animal to cope. This may be as true of amphetamine stereotypies as it is of environment-induced stereotypies (Jones *et al.*, 1989; Mittleman *et al.*, 1991). The evidence is reviewed by Mason (1991a), and discussed in Chapters 5, 6 and 7.

DIFFERENCES BETWEEN INDIVIDUALS

The role of age and experience in the development of idiosyncratic stereotypies had already been discussed. However, they cannot account for all the individual differences seen. Individuals often differ in their stereotypies, even when their past and present circumstances are very similar. For example, this is true of the environment-induced stereotypies of pigs (Wood-Gush *et al.*, 1983; Schlichting and Smidt, 1984; Cronin, 1985; Appleby and Lawrence, 1987), bank voles (Ödberg, 1978, 1986), pigeons (Palya and Zacny, 1980), rats (Staddon and Ayres, 1975), calves (Wiepkema, 1984, 1987), and fennec foxes Ödberg, 1984b). This is true of stimulant-induced stereotypies, too (e.g. Lyon and Robbins, 1975; Benus *et al.*, 1991; Terlouw *et al.*, 1993 a,b). Animals of the same age and experience may even differ in the extent to which their stereotypies become 'established'. For example, although most canaries that develop stereotypies through deprivation of nesting-material lose the behaviour once suitable substrates are provided, some individuals do not, and retain their stereotypies (Hinde, 1958).

Some of these individual differences could be due to differences in motivation; some animals may find the environment more eliciting than others (e.g. individual pigs have different levels of appetite; Fraser, 1980, p. 180). Rhesus monkeys reared with an artificial mother differ in the amount of time they spend body-rocking; Mason and Berkson (1975) suggest that this is because they differ in how much proprioceptive and kinaesthetic stimulation they require. Some individual differences may also be the product of other factors.

Individuals differ in their tendency to respond to environmental stimuli with active responses rather than in a more inactive way, in their tendency to persist with behavioural responses, and in their general propensities to develop inflexible behavioural routines. These differences are all likely to influence the tendency of an individual to develop stereotypies (reviewed by Mason, 1991b). See also Chapters 3 and 7.

Studying Stereotypies

Research methods must take into account that stereotypies are not a neat, homogeneous, self-contained group of behaviour patterns. As there is a subjective element in identifying stereotypies, the criteria used must be stated so that observations are repeatable. And as different stereotypies are influenced by different factors, and vary so much in their own characteristics, they and their circumstances must be thoroughly described.

GENERAL ETHOLOGICAL TECHNIQUES

General ethological techniques are important in studying stereotypies. For example, the reliability of data collection should be checked for intra- and inter-observer error (e.g. Martin and Bateson, 1986, pp. 88–89). Also, sufficient data should be collected to ensure the validity of the results. One way of checking for reliability is to divide the data randomly into two halves and calculate the correlations between the two sets of values. If the correlations are high, the data are internally consistent and reliable (Martin and Bateson, 1986, p. 32).

Large sample sizes allow for a more accurate assessment of population means, and will give a clearer picture of how individuals vary according to age and sex. Stereotypies should also be observed over long periods of time, to accurately record diurnal and seasonal rhythms, for example as are found in mink (Mason, 1992). However, a sample size is only increased for statistical purposes if the data points are independent of each other, representing different, and preferably unrelated, individuals. The assumption of independence is frequently violated in behavioural research. For example, repeated observations of the same subject are not independent; collecting the details of 100 stereotypic bouts in two animals obviously does not provide the same information as observing ten bouts in each of 20 animals. Similarly, littermates are not necessarily independent (Martin and Bateson, 1986, pp. 28–30). A link between stereotypy and corticosteroid levels, for example, would tell us little if the data were collected from only two litters; the relationship could be due to differences between the families, rather than to genuine, behaviour-related differences between the individuals.

Measurements specifically important for stereotypies

Is it a stereotypy?

To justify calling something a stereotypy, researchers must explicitly quantify 'repetitiveness' and 'rigidity', and say why they think a behaviour has no apparent function.

To quantify repetitiveness or intensity, one could measure the mean or maximum number of cycles per bout, and the number of bouts per day. Another measure is to assess the ease with which repetition is broken by an external stimulus, such as a loud tone (Fentress, 1976; Robbins and Sahakian, 1981).

Quantifying variability means estimating the extent to which a behaviour pattern is similar on different occasions. I used a simple method to estimate the variability of mink stereotypies. I described the pattern of movements they made in each 30-second observation period, and then calculated the proportion of total observations for each animal in which the sequences of movements were completely unvarying (Mason, 1992). Sequential analysis can also be used to assess the order within a sequence, and the probability of transition between different elements (e.g. Cronin, 1985; Martin and Bateson, 1986, pp. 63–65). Alternatively, informational redundancy can be used to quantify the predictability of behavioural sequences (Stolba et al., 1983), although this method is not infallible. Fentress and Stilwell (1973) used information measures to assess the structure of face-grooming sequences in mice. They found that although the data analysis did indicate some sequential structure, it did not convey the degree of order apparent to the human observer. Analysing longer sequences, to get results that they intuitively felt would be more representative, would have been more computationally cumbersome and would also have obscured details. Consideration of the variability of behaviour patterns may lead to their classification into different subgroups. For example, Cronin and Wiepkema (1984) collected detailed observations of the movements made by tethered sows, and found that in addition to unambiguous stereotypies made up of sequences of repeated cycles of elements, sows performed behaviour that was more variable. Variable sequences were sometimes seen when an animal was changing from one stereotypic sequence to another; these they called 'transitional stereotypies'. Behaviour patterns that they termed 'loose stereotypies' were also seen. These were cycles of fixed, predictable subroutines together with variable, unpredictable elements.

When assessing the variability of behaviour, results may depend on the timescale considered. The timescale used should, therefore, always be made explicit. For example, within a bout, an animal may vary the sequences of its movements. However, when several bouts are observed, it may become apparent that on a more gross level the animal is very predictable. This is true of mink where even those animals with variable stereotypic sequences were usually very predictable in the exact movements that they employed, the parts of a sequence that they were most likely to vary, and in subtle aspects of 'style' such as their speed of performance, whether vocalization was involved, and idiosyncratic traits such as the amplitude or fluidity of their movements. Over this longer timescale, the animals' behaviour was predictable not only from bout to bout, but from day to day and even year to year (Mason, 1992).

The degree to which a stereotypy is fixed in form, and its repetitiveness or duration, may be quite independent characteristics; a stereotypy may be repetitive yet varying, or be performed infrequently yet be very predictable in form (e.g. Mason and Turner, 1993). Thus studies like that of Rushen (1984) and Terlouw et al. (1991a), that focus on measures of duration with only qualitative judgements about the degree of fixity, do not provide full information about the stereotypy. Both aspects should be measured to describe the behaviour pattern fully, and to justify its classification as a stereotypy.

In addition, researchers must stipulate why it is they judge the behaviour to be without goal or function. To what degree does it quantitatively differ from normal behaviour? For example, the inappropriate nature of SIP in rats is revealed by their ingestion of bitter quinine (e.g. Roper, 1983), and that of amphetamine stereotypy by the way in which rats will repeatedly press operant levers without taking any of the rewarding food pellets (Evenden and Robbins, 1983). In some cases attributing lack of function to a behaviour is more controversial. Terlouw et al. (1991b, 1993b) consider drinking in sows to be a stereotypy, as some animals consume as much as 54 litres of water a day. However, as some pigs are consuming a mere 6 litres a day, and there is a range of values between these two extremes, at some point the volumes must cease to be excessive and start to fall within the normal metabolic requirements of the animal. Different workers may disagree as to when a particular behaviour pattern has no obvious goal or function (e.g. Cronin and Wiepkema, 1984, did not class anything as a stereotypy until feeding and drinking were over, in their sows), but if the reasons for these beliefs are presented then at least then the matter is open to debate.

What type of stereotypy is it?

Having identified a behaviour as stereotypic, what else do we need to know? The heterogeneity of stereotypies requires that their morphology is always described, along with certain other characteristics. The following information should be recorded to assess which other stereotypies a new example most resembles:

MORPHOLOGY

As well as quantitative data such as bout length and redundancy measures, a detailed description of the stereotypy should always be provided. Using video may help here (e.g. Jensen *et al.* 1986, pp. 34–37). Descriptions should include the parts of the body and the substrates involved (e.g. Cronin and Wiepkema 1984), the degree of individual variation, and the similarities and differences between the stereotypy and other, normal, behaviour patterns. This comparison should include measures of rapidity (Dantzer, 1986). The 'local rate' of a behaviour pattern is the number of component acts per unit time spent performing the activity (Roper, 1984). This measure would also provide a simple and informative way of quantifying intensity (Martin and Bateson, 1986, p. 46), for example to compare bouts of stereotypy performed at different times. Descriptions could also extend to facial expressions while performing the behaviour and other impressions gained of the associated emotions or arousal level. To enable other researchers to replicate the study, one should also stipulate the criteria used to distinguish between a break within a bout, and between one bout and the next (e.g. Jensen *et al.*, 1986, p. 60; Rushen *et al.*, 1990).

If a variety of different forms are observed, this raises the issue of how they should be combined, if at all. Stolba *et al.* (1983), Cronin and Wiepkema (1984), Appleby and Lawrence (1987) and von Borell and Hurnik (1991) combined stereotypies of different forms into one category, yet other researchers may have distinguished between them on the grounds of their behavioural content.

Some studies use rating scales to quantify the nature of animals' stereotypies. This is a particularly common technique for stimulant-induced stereotypies, and the scale is based on the behaviour typical of greater and greater doses of the drug (Robbins and Sahakian, 1981; Benus *et al.*, 1991). Although relatively simple to implement, this method has a number of problems that make it inadvisable for researchers asking ethological questions. One problem is that the scale combines all forms of stereotypy; for example, it would not yield separate data on locomotory and oral stereotypies. Second, the rating scales do not always conform closely to the consensual definition of stereotypy; that they are based on changes in behaviour with increasing dose of a drug does not mean that they therefore chart greater and greater levels of stereotypy as we would define it. In the scale used by Benus *et al.* (1991), for example, we might well question the ranking of 'stereotyped behaviour pattern in a particular place' as less stereotyped than 'stereotyped behaviour pattern in a particular place with intermittent gnawing and licking'. Third, it provides no information on bout length and local rate and so on. Additional problems are considered by Robbins and Sahakian (1981). The scales used may not reliably reflect dose-related phenomena, since oral stereotypies do not always supplant locomotory ones in the rat; the reverse may sometimes occur, particularly when the agent used is apomorphine. Also, the scales do not adequately assess changes in behaviour considered normal; for example, stereotyped switching between operant levers would not be recorded at all.

WHEN IS IT PERFORMED?
Data should be collected on the temporal pattern of the stereotypy, such as the frequency of bout initiation, and the contexts in which this occurs. Data should also describe changes in its nature or form when performed in different circumstances. For example, detailed (unpublished) data collected by de Jonge on the stereotypies of mink have shown some stereotypies to increase more than others as feeding time approaches.

CURRENT AGE AND CIRCUMSTANCES OF THE SUBJECT
The animal's age and stage of development at the time of stereotypy performance should be recorded, along with a description of the physical and social environment in which it is housed, and its feeding regime.

STAGE OF DEVELOPMENT OF THE STEREOTYPY
Information on the development of a stereotypy should include the length of time for which the animal has performed the behaviour, whether it is performed in situations other than the original eliciting one; and whether the stereotypy is readily modifiable (e.g. reduced by environmental enrichment or interrupted by, say, a loud noise). Note that we cannot assume that all the stereotypies of one subject are all of same type (e.g. Dodman *et al.*, 1987); these data should be collected separately for different forms of an individual's stereotypy.

PAST HISTORY OF THE SUBJECT
It is important to record, finally, the first circumstances in which the stereotypy was performed, if known; the early history of the animal (e.g. whether captive-bred or wild-caught), and the age at which it was removed from its mother and siblings. In addition, one should consider the influence that the behaviour of conspecifics may have had on the subject's behaviour (as models for imitation, for example).

Conclusions

Because 'stereotypy' is not a category with a clearly defined border, the stereotypies described by some authors may be classed differently by others. However, this may not be a substantial problem as long as all researchers clearly describe, qualitatively and quantitatively, each behaviour pattern they so define.

There are many differences between the various forms of stereotypy.

Different stereotypies may share some characteristics but not others, and, furthermore, these characteristics may also be shared by behaviour patterns outside the group such as habits, tics and daily routines (see e.g. Mason and Turner, 1993). Such classification problems are not unique to stereotypies. For example, the term 'play' has been used to cover a broad range of juvenile behaviour because a consensual definition has proved so elusive (Martin and Caro, 1985); no one single characteristic defines play, and the behaviour is still poorly understood. Likewise, arousal is sometimes presented as a unitary concept, when it more accurately should be considered as a number of different phenomena (Robbins and Everitt, 1987); and self-mutilation in human psychological conditions has different causal bases and multiple possible functions (Carr, 1977). As with these cases of heterogeneous phenomena, we should not overlook the special or unique properties of each case of stereotypy in our desire to classify and hence make order out of apparent chaos. Nor should we fall into the trap of trying to explain all aspects of stereotypies in terms of welfare (see Mason, 1991b). This emphasizes the importance of collecting full and detailed information. To date, stereotypies have been measured in many different ways. Some studies have assessed duration and not variability, while others have done the opposite, and some studies have distinguished between stereotypies of different forms where others have combined them. This has made it difficult to compare results, and has been a major impediment to understanding (Mason, 1991a). Future studies should approach stereotypies in a multi-dimensional way, describing and measuring each of the different properties. Then, by pooling detailed information from stereotypies that appear genuinely equivalent in their source behaviour patterns and similar in their current morphology, timing and stage of development, reliable general theories and classificatory schemes for stereotypies can continue to emerge, as our understanding of the variety of different forms increases.

Acknowledgements

With thanks to Michelle Turner, who provided me with information on human cases.

References

Anderson, M.C. and Shettleworth, S.J. (1977) Behavioural adaptation to fixed-interval and fixed-time food delivery in golden hamsters. *Journal of the Experimental Analysis of Behaviour* 25, 33–49.

Appleby, M.C. and Lawrence, A.B. (1987) Food restriction as a cause of stereotypic behaviour in tethered sows. *Animal Production* 45, 103–110.

Appleby, M.C., Lawrence, A.B. and Illius, A.W. (1989) Influence of neighbours on stereotypic behaviour of tethered sows. *Applied Animal Behaviour Science* 24, 137–146.

Baron-Cohen, S. (1989) Do autistic children have obsessions and compulsions? *British Journal of Clinical Psychology* 28, 193–200.

Bateson, P.P.G. (1986) Functional approaches to behaviour and development. In: Else, J.G. and Lee, P.C. (eds), *Primate Ontogeny, Cognition and Social Behaviour*. Cambridge University Press, Cambridge, pp. 183–192.

Benus, R.F., Bohus, B., Koolhaas, J.M. and van Oortmerssen, G.A. (1991) Behavioural differences between artificially selected aggressive and non-aggressive mice: response to apomorphine. *Behavioural Brain Research* 43, 203–208.

Berkson, G. (1967) Abnormal stereotyped motor acts. In: Zubin, J. and Hunt, H.F. (eds), *Comparative Psychopathology – Animal and Human*. Grune and Stratton, New York, pp. 76–94.

Berkson, G. (1968) Development of abnormal stereotyped behaviours. *Developmental Psychobiology* 1, 118–132.

Berkson, G. (1983) Repetitive stereotyped behaviours. *American Journal of Mental Deficiency* 88, 239–246.

Berkson, G. and Mason, W.A. (1964) Stereotyped behaviors of chimpanzees: relation to general arousal and alternative activities. *Perceptual and Motor Skills* 19, 635–652.

Berkson, G., Mason, W.A. and Saxon, S.V. (1963) Situation and stimulus effects on stereotyped behaviors of chimpanzees. *Journal of Comparative Physiology and Psychology* 56, 786–792.

Berlyne, D.E. (1960) *Conflict, Arousal and Curiosity*. McGraw-Hill, New York.

Bernstein, S. and Mason, W.A. (1962) The effects of age and stimulus conditions on the emotional responses of rhesus monkeys: responses to complex stimuli. *Journal of Genetic Psychology* 101, 279–298.

Blozovski, D. (1986) L'hippocampe et le comportement. *La Recherche* 17, 330–338.

Bolles, R.C. (1970) Species-specific defense reactions and avoidance learning. *Psychological Review* 77, 32–48.

Boucher, J. (1977) Alternation and sequencing behaviour, and response to novelty, in autistic children. *Journal of Child Psychology and Psychiatry* 18, 67–72.

Bowlby, J. (1988) *A Secure Base: Clinical Applications of Attachment Theory*. Routledge, London.

Broom, D.M. (1983) Stereotypies as animal welfare indicators. In: Smidt, D. (ed.), *Indicators Relevant to Farm Animal Welfare*. Martinus Nijhoff, The Hague, pp. 81–87.

Cabib, S., Puglisis-Allegra, S. and Oliverio, A. (1985) A genetic analysis of stereotypy in the mouse. *Behavioral and Neural Biology* 44, 239–248.

Carpenter, R.H.S. (1989) *Neurophysiology*, 2nd edition. Edward Arnold, London.

Carr, E.G. (1977) The motivation of self-injurious behaviour: a review of some hypotheses. *Psychological Bulletin* 84, 800–816.

Cronin, G. (1985) The development and significance of abnormal stereotyped behaviours in tethered sows. PhD Thesis, Agricultural University of Wageningen, The Netherlands.

Cronin, G.M. and Wiepkema, P.R. (1984) An analysis of stereotyped behaviour in tethered sows. *Annales de Recherches Vétérinaires* 15, 263–270.

Cronin, G.M., Wiepkema, P.R. and Hofstède, G.J. (1984) The development of stereotypes in tethered sows. In: Unshelm, J., van Putten, G. and Keeb, K. (eds), *Proceedings of the International Congress on Applied Ethology in Farm Animals*. KTBL, Darmstadt, pp. 97–100.

Cronin, G.M., Wiepkema, P.R. and van Ree, J.M. (1985a) Endogenous opioids are involved in abnormal stereotyped behaviours of tethered sows. *Neuropeptides* 6, 527–530.

Cronin, G.M., Wiepkema, P.R. and Mekking, P. (1985b) Stereotypy performance characteristics of tethered sows in a commercial herd and the relationship to sow welfare and productivity. In: Cronin, G. The development and significance of abnormal stereotyped behaviours in tethered sows. PhD Thesis, Agricultural University of Wageningen, The Netherlands.

Cronin, G.M. van Tartwijk, J.M.F.M., van der Hel, W. and Verstegen, M.W.A. (1986) The influence of degree of adaptation to tether-housing by sows in relation to behaviour and energy metabolism. *Animal Production* 42, 257–268.

Dantzer, R. (1986) Behavioral, physiological, and functional aspects of stereotyped behavior: a review and a re-interpretation. *Journal of Animal Science* 62, 1776–1786.

Dantzer, R. (1991) Stress, stereotypies and welfare. *Behavioural Processes* 25, 95–102.

Dantzer, R. and Mormède, P. (1981) Pituitary–adrenal consequences of adjunctive activities in pigs. *Hormones and Behaviour* 15, 386–395.

Dantzer, R. and Mormède, P. (1983) De-arousal properties of stereotyped behaviour: evidence from pituitary-adrenal correlates in pigs. *Applied Animal Ethology* 10, 233–244.

Davenport, R.K. and Menzel, E.W. (1963) Stereotyped behaviour of the infant chimpanzee. *Archives of General Psychiatry* 8, 99–104.

Dawkins, M. (1986) *Unravelling Animal Behaviour*. Longman, Harlow, Essex.

de Jonge, G., Carlstead, K. and Wiepkema, P.R. (1986) *The Welfare of Ranch Mink*. (Translated from Dutch.) COVP Issue No. 08, Het Spelderholt, Beekbergen, The Netherlands.

Dellmeier, G.R. (1989) Motivation in relation to the welfare of enclosed livestock. *Applied Animal Behaviour Science* 22, 129–138.

Devenport, L.D., Devenport, J.A. and Holloway, F.A. (1981) Reward-induced stereotypy: modulation by the hippocampus. *Science* 212, 1288–1289.

Dodman, N.H., Shuster, L., Court, M.H. and Dixon, R. (1987) Investigation into the use of narcotic antagonists in the treatment of a stereotypic behaviour pattern (crib-biting) in the horse. *American Journal of Veterinary Research* 48, 311–319.

Draper, W.A. and Bernstein, I.S. (1963) Stereotyped behavior and cage size. *Perceptual and Motor Skills* 16, 231–234.

Duncan, I.J.H. (1970) Frustration in the fowl. In: Freeman, B.M. and Gordon, R.G. (eds), *Aspects of Poultry Behaviour*. British Poultry Science, Edinburgh, pp. 15–31.

Duncan, I.J.H. and Wood-Gush, D.G.M. (1972) Thwarting of feeding behaviour in the domestic fowl. *Animal Behaviour* 20, 444–451.

Duncan, I.J.H. and Wood-Gush, D.G.M. (1974) The effect of a rauwolfia tranquillizer on stereotyped movements in frustrated domestic fowl. *Applied Animal Ethology* 1, 67–76.

Eichler, A.J., Antelman, S.M. and Black, C.A. (1980) Amphetamine stereotypy is not a homologous phenomenon: sniffing and licking show distinct profiles of sensitization and tolerance. *Psychopharmacology* 6, 287–290.

Einon, D.F. and Sahakian, B.J. (1979) Environmentally-induced differences in susceptibility of rats to CNS stimulants and CNS depressants: evidence against a unitary explanation. *Psychopharmacology* 61, 299–307.

Ellinwood, E.H. and Kilbey, M.M. (1975) Amphetamine stereotypy: the influence of environmental factors and prepotent behavioral patterns on its topography and development. *Biological Psychiatry* 10, 3–16.

Evenden, J.L. and Robbins, T.W. (1983) Increased response switching, perseveration and perseverative switching following d-amphetamine in the rat. *Psychopharmacology* 80, 67–73.

Falk, J.L. (1971) The nature and determinants of adjunctive behavior. *Physiology and Behaviour* 6, 577–588.

Feldman, R.S. and Green, K.F. (1967) Antecedents to behaviour fixations. *Psychological Review* 74, 250–271.

Fentress, J.C. (1973) Specific and non-specific factors in the causation of behaviour. In: Bateson, P.P.G. and Klopfer, P.H. (eds), *Perspectives in Ethology*. Plenum Press, New York, pp. 155–224.

Fentress, J.C. (1976) Dynamic boundaries of patterned behaviour: interaction and self-organisation. In: Bateson, P.P.G. and Hinde, R.A. (eds), *Growing Points in Ethology*. Cambridge University Press, Cambridge, pp. 135–167.

Fentress, J.C. (1977) The tonic hypothesis and the patterning of behavior. *Annals – New York Academy of Sciences* 290, 370–395.

Fentress, J.C. and Stilwell, F.P. (1973) Grammar of a movement sequence in inbred mice. *Nature* 244, 52–53.

Forehand, R. and Baumeister, A.A. (1971) Stereotyped body-rocking as a function of situation, IQ, and time. *Journal of Clinical Psychology* 27, 324–326.

Fox, M.W. (1965) Environmental factors influencing stereotyped and allelomimetic behavior in animals. *Laboratory Animal Care* 15, 363–370.

Fox, M.W. (1971) Psychopathology in man and lower animals. *Journal of the American Veterinary Medical Association* 159, 66–77.

Frances, S. (1966) An ethological study of mentally retarded individuals and normal individuals. Unpublished PhD Thesis, University of Cambridge.

Fraser, A.F. (1968) Behavior disorders in domestic animals. In: Fox, M.W. (ed.), *Abnormal Behavior in Animals*, Saunders, London, pp. 179–187.

Fraser, A.F. (1980) *Farm Animal Behaviour*, 2nd edition. Baillière-Tindall, London.

Fraser, A.F. and Broom, D.M. (1990) *Farm Animal Behaviour and Welfare*. Bailliere-Tindall, London.

Fray, P.J., Sahakian, B.J., Robbins, T.W., Koob, G.F. and Iverson, S.D. (1980) An observational method for quantifying the behavioural effects of dopamine agonists: contrasting effects of d-amphetamine and apomorphine. *Psychopharmacology* 69, 253–259.

Frith, C.D. and Done, D.J. (1990) Stereotyped behaviour in madness and in health. In: Cooper, S.J. and Dourish, C.T. (eds), *Neurobiology of Stereotyped Behaviour*. Clarendon, Oxford, pp. 233–259.

Fry, J.P., Sharman, D.F. and Stephens, D.B. (1981) Cerebral dopamine, apomorphine and oral activity in the neonatal pig. *Journal of Veterinary Pharmacology and Therapeutics* 4, 193.

Gallagher, R.J. and Berkson, G. (1986) Effect of intervention techniques in reducing hand-gazing in young severely disabled children. *American Journal of Mental Deficiency* 91, 170–177.

Garrigues, P., Peterson, A.F., de Roquefeuil, G. and Gourdon, N. (1982) On behavioural similarities between animal and human stereotypies, with a note on a preliminary analysis of stereotypic movements in psychotic children. Abstract, *Applied Animal Ethology* 9, 197.

Groothuis, T. (1989) On the ontogeny of display behaviour in the black-headed gull: I. The gradual emergence of the adult forms. *Behaviour* 109, 76–124.

Harlow, H.J. and Zimmerman, R.R. (1959) Affectional responses in the infant monkey. *Science* 130, 421–432.

Hediger, H. (1950) *Wild Animals in Captivity*, 2nd edition. Butterworth Scientific Publications, London.

Hinde, R.A. (1958) The nest-building behaviour of domesticated canaries. *Proceedings of the Zoological Society of London* 131, 1–48.

Hinde, R.A. (1962) The relevance of animal studies to human neurotic disorders In: Richter, D., Tanner, J.M., Lord Taylor and Zangwill, O.L. (eds), *Aspects of Physchiatric Research*. Oxford University Press, London, pp. 240–261.

Hinde, R.A. (1970) *Animal Behaviour*, 2nd edition. McGraw-Hill, New York.

Houpt, K.A. (1987) Abnormal behavior. In: Price, E.O. (ed.) *The Veterinary Clinics of North America – Food Animal Practice, Vol. 3 (2) – Farm Animal Behavior*. W.B. Saunders, Philadelphia, pp. 357–368.

Hudson, A.J. (1969) Perseveration. *Brain* 91, 571–582.

Hutchinson, R.R. (1977) By-products of aversive conditioning. In: Honig, W.K. and Staddon, J.E.R. (eds), *The Handbook of Operant Behaviour*. Prentice Hall, Englewood Cliffs, New Jersey, pp. 415–430.

Hutt, C. and Hutt, S.J. (1965) The effects of environmental complexity on the stereotyped behaviour of children *Animal Behaviour* 13, 1–4.

Jensen, P., Algers, B. and Ekesbo, I. (1986) *Methods of Sampling and Analysis of Data in Farm Animal Ethology*. Birkhauser Verlag, Basle.

Jones, G.H., Mittleman, G. and Robbins, T.W. (1989) Attenuation of amphetamine-stereotypy by mesostriatal dopamine depletion enhances plasma corticosterone: implications for stereotypy as a coping response. *Behavioural and Neural Biology* 51, 80–91.

Jones, I. (1965) Observations of schizophrenic stereotypies. *Comprehensive Psychiatry* 6, 323–335.

Keiper, R.R. (1969) Causal factors of stereotypies in caged birds. *Animal Behaviour* 17, 114–119.

Keiper, R.R. (1970) Studies of stereotypy function in the canary (*Serinus canarius*). *Animal Behaviour* 18, 353–357.

Kennes, D. and Ödberg, F.O. (1987) Developmental study of the effect of haloperidol and naloxone on captivity-induced stereotypies. Abstract, *Applied Animal Behaviour Science* 17, 379.

Kennes, D., Ödberg, F.O., Bouquet, Y. and de Rycke, P.H. (1988) Changes in naloxone and haloperidol effects during the development of captivity-induced jumping stereotypy in bank voles. *European Journal of Pharmacology* 153, 19–24.

Kiley-Worthington, M. (1977) *Behavioural Problems of Farm Animals*. Oriel Press, London.

Kiley-Worthington, M. (1983) Stereotypes in horses. *Equine Practice* 5, 34–40.
Koegel, R.L. and Covert, A. (1972) The relationship of self-stimulation to learning in autistic children. *Journal of Applied Behavior Analysis* 5, 381–387.
Kravitz, H., Rosenthal, V., Teplitz, Z., Murphy, J.B. and Lesser, R.E. (1960) A study of head-banging in infants and children. *Diseases of the Nervous System* 21, 203–208.
Lashley, K.S. (1917) The accuracy of movement in the absence of excitation from the moving organ. *American Journal of Physiology* 43, 169–194.
Lashley, K.S. (1921) Studies of cerebral function in learning. II: the effects of long-continued practice upon cerebral localization. *Journal of Comparative Psychology* 1, 453–468.
Lashley, K.S. (1951) The problem of serial order in behaviour. In: Jeffress, L.A. (ed.), *Cerebral Mechanisms in Behaviour*. John Wiley, New York, pp. 112–136.
Levy, D.M. (1944) On the problem of movement restraint. *American Journal of Orthopsychiatry* 14, 644–677.
Lovaas, O.I., Litrovik, A. and Mann, R. (1971) Response latencies to auditory stimuli in autistic children engaged in self-stimulatory behaviour. *Behaviour Research and Therapy* 9, 39–49.
Luria, A.R. (1973) *The Working Brain*. Allen Lane, The Penguin Press, London.
Lyon, M. and Robbins, T. (1975) The action of CNS stimulant drugs: a general theory concerning amphetamine effects. In: Essman, W. and Valzelli, L. (eds), *Current Developments in Psychopharmacology, Vol. 2*. Spectrum Publications, New York, pp. 81–163.
MacLennan, A.J. and Maier, S.F. (1983) Coping and the stress-induced potentiation of stimulant stereotypy in the rat. *Science* 219, 1091–1093.
Marks, I.M. (1987) *Fears, Phobias and Rituals*. Oxford University Press, New York.
Martin, P. and Bateson, P.P.G. (1986) *Measuring Behaviour – An Introductory Guide*. Cambridge University Press, Cambridge.
Martin, P. and Bateson, P.P.G. (1988) Behavioural development in the cat. In: Turner, D.C. and Bateson, P.P.G. (eds), *The Domestic Cat*. Cambridge University Press, Cambridge, pp. 9–22.
Martin, P. and Caro, T.M. (1985) On the functions of play and its role in behavioural development. *Advances in the Study of Behaviour* 15, 59–102.
Martiniuk, R.G. (1976) *Information Processing in Motor Skills*. Holt, Rinehart and Winston, New York.
Mason, G.J. (1991a) Stereotypies: a critical review. *Animal Behaviour* 41, 1015–1037.
Mason, G.J. (1991b) Stereotypies and suffering. *Behavioural Processes* 25, 103–115.
Mason, G.J. (1992) Individual differences in the stereotypies of caged mink. Unpublished Ph D Thesis, University of Cambridge.
Mason, G.J. and Turner, M.A. (1993) Mechanisms involved in the development and control of stereotypies. In: Bateson, P.P.G. and Klopfer, P.H. (eds), *Perspectives in Ethology 10*. Plenum Press, New York (in press).
Mason, W.A. and Berkson, G. (1975) Effects of maternal mobility on the development of rocking and other behaviors in rhesus monkeys: a study with artificial mothers. *Developmental Psychobiology* 8, 197–211.
Mason, W.A. and Green, P.C. (1962) The effects of social restriction on the behavior of rhesus monkeys. IV: responses to a novel environment and to an alien species. *Journal of Comparative Physiology and Psychology* 55, 363–368.

McFarland, D.J. (1985) *Animal Behaviour*. Longman Scientific and Technical, Harlow.

McMahon, F.B. and McMahon, J.W. (1983) *Abnormal Behaviour: Psychology's View*. Dorsey Press, Homewood, Illinois.

Meyer-Holzapfel, M. (1968) Abnormal behaviour in zoo animals. In: Fox, M.W. (ed.), *Abnormal Behavior in Animals*. Saunders, London, pp. 476–503.

Miller, G.A., Galenter, E. and Pribram, K.H. (1960) *Plans and Structure of Behaviour*. Holt, Rinehart and Winston, New York.

Mittleman, G. and Valenstein, E.S. (1985) Individual differences in non-regulatory ingestive behaviors and catecholamine systems. *Brain Research* 348, 112–117.

Mittleman, G., Castaneda, E., Robinson, T.E. and Valenstein, E.S. (1986) The propensity for nonregulatory ingestive behavior is related to differences in dopamine systems: behavioral and biochemical evidence. *Behavioural Neuroscience* 100, 213–222.

Mittleman, G., Jones, G.H. and Robbins, T.W. (1991) Sensitization of amphetamine stereotypy reduces plasma corticosterone: implications for stereotypy as a coping response. *Behavioral and Neural Biology* 56, 170–183.

Morgan, M.J. (1973) Effects of post-weaning environment on learning in the rat. *Animal Behaviour* 21, 429–442.

Morris, D. (1964) The response of animals to a restricted environment. *Symposium of the Zoological Society of London* 13, 99–118.

Morris, D. (1966) The rigidification of behaviour. *Philosophical Transactions of the Royal Society of London, Series B* 251, 327–330.

Nielsen, B.L., de Rosa, G., Whittemore, J., Terlouw, E.M.C. and Lawrence, A.B. (1993) Individual differences in the behaviour of sows subjected to an intermittent feeding schedule. *Behavioural Processes* (in press).

Ödberg, F. (1978) Abnormal behaviours (stereotypies), Introduction to the Round Table. In: *Proceedings of the First World Congress of Ethology Applied to Zootechnics, Madrid*, Editorial Garsi, Industrias Graficas Espana, Madrid, pp. 475–480.

Ödberg, F. (1984a). Neurochemical correlates of stereotyped behaviour in voles. Abstract. *Applied Animal Behaviour Science* 13, 168.

Ödberg, F. (1984b) The altering of stereotypy levels: an example with captive fennecs (*F. zerda*). In: Unshelm, J., van Putten, G. and Zeeb, K. (eds), *Proceedings of the International Congress on Applied Ethology in Farm Animals*. KTBL, Darmstadt, pp. 299–301.

Ödberg, F. (1986) The jumping stereotypy in the bank vole (*Clethrionomys glareolus*). *Biology of Behavior* 11, 130–143.

Ödberg, F. (1987) Behavioural responses to stress in farm animals. In: Wiepkema, P.R. and van Adrichem, P.W.M. (eds), *The Biology of Stress in Farm Animals: An Integrative Approach*. Martinus Nijhoff, Dordrecht, pp. 135–149.

Ödberg, F. (1989) Behavioural coping in chronic stress conditions. In: Blanchard, R.J., Brain, P.F., Blanchard, D.C. and Parmigiani, S. (eds), *Ethoexperimental Approaches to the Study of Behavior*. Kluwer Academic Press, Dordrecht, pp. 229–238.

Palya, W.L. and Zacny, J.P. (1980) Stereotyped adjunctive pecking by caged pigeons. *Animal Learning and Behaviour* 8, 293–303.

Pribram, H. and McGuiness, D. (1975) Arousal, activation and effort in the control of attention. *Psychological Review* 82, 116–149.

Prior, M. and MacMillan, M.B. (1973) Maintenance of sameness in children with Kanner's syndrome. *Journal of Autism and Childhood Schizophrenia* 3, 154–167.

Rachman, S. (1985) Obsessional-compulsive disorder. In: Bradley, B.P. and Thompson, C. (eds), *Psychological Applications in Psychiatry*. J. Wiley and Sons Ltd, Chichester, pp. 7–39.

Ridley, M. (1986) *Evolution and Classification – The Reformation of Cladism*. Longman, London.

Ridley, R.M. and Baker, H.F. (1982) Stereotypy in monkeys and humans. *Psychological Medicine* 12, 61–72.

Robbins, T.W. (1976) Relationship between reward-enhancing and stereotypical effects of psychomotor stimulant drugs. *Nature* 264, 57–59.

Robbins, T.W. (1982) Stereotypies: addictions or fragmented actions? *Bulletin of the British Psychological Society* 35, 297–300.

Robbins, T.W. (1991) Cognitive deficits in schizophrenia and Parkinson's disease: neural basis and the role of dopamine. In: Willner, P. and Scheel-Kruger, J. (eds), *The Mesolimbic Dopamine System: From Motivation to Action*. J. Wiley and Sons Ltd, Chichester, pp. 497–528.

Robbins, T.W. and Everitt, B.J. (1982) Functional studies of the central catecholamines. *International Review of Neurobiology* 23, 303–365.

Robbins, T.W. and Everitt, B.J. (1987) Psychopharmacological studies of arousal and attention. In: Stahl, S.M. Iverson, S.D. and Goodman, E.C. (eds), *Cognitive Neurochemistry*. Oxford University Press, London, pp. 135–170.

Robbins, T.W. and Sahakian, B.J. (1981) Behavioural and neurochemical determinants of drug-induced stereotypy. In: Clifford, R.F. (ed.), *Metabolic Disorders of the Nervous System*. Pitman, London, pp. 244–291.

Robbins, T.W. and Sahakian, B.J. (1983) Behavioral effects of psychomotor stimulant drugs: clinical and neuropsychological implications. In: Creese I. (ed.), *Stimulants: Neurochemical, Behavioral and Clinical Perspectives*. Raven Press, New York, pp. 301–338.

Robbins, T.W., Jones, G.H. and Sahakian, B.J. (1989) Central stimulants, transmitters and attentional disorder: a perspective from animal studies. In: Sagrolden, T. and Archer, T. (eds), *Attention Deficit Disorders*. Lawrence Erlbaum Associates, New Jersey, pp. 199–222.

Robbins, T.W., Mittleman, G., O'Brien, J. and Winn, P. (1990) The neuropsychological significance of stereotypy induced by stimulant drugs. In: Cooper, S.J. and Dourish, C.T. (eds), *The Neurobiology of Stereotyped Behaviour*. Clarendon Press, Oxford, pp. 25–63.

Roper, T.J. (1980) Changes in the rate of schedule-induced behaviour in rats as a function of fixed-interval schedule. *Quarterly Journal of Experimental Psychology* 32, 159–170.

Roper, T.J. (1983) Schedule-induced behaviour. In: Mellgren, R.S. (ed.) *Animal Cognition and Behaviour*. North-Holland Publishing Company, Amsterdam, pp. 127–164.

Roper, T.J. (1984) Response of thirsty rats to the absence of water: frustration, disinhibition or compensation? *Animal Behaviour* 32, 1225–1235.

Rushen, J. (1984) Stereotyped behaviour, adjunctive drinking and the feeding periods of tethered sows. *Animal Behaviour* 32, 1059–1067.

Rushen, J. (1985) Stereotypies, aggression and the feeding schedules of tethered sows. *Applied Animal Behaviour Science* 14, 137–147.

Rushen, J., Passillé, A.M. de and Schouten, W. (1990) Stereotyped behaviour, endogenous opioids and post-feeding hypoalgesia in pigs. *Physiology and Behaviour* 48, 91–96.

Rutter, M. (1981) *Maternal Deprivation Reassessed*, 2nd edition. Penguin, Harmondsworth, Middlesex.

Sacks, O. (1990) *Awakenings*, revised edition. Picador, Pan Books, London.

Sahakian, B.J. and Robbins, T.W. (1975) The effects of test environment and rearing conditions on amphetamine-induced stereotypy in the guinea-pig. *Psychopharmacologia* 45, 115–117.

Sahakian, B.J., Robbins, T.W., Morgan, M.J. and Iverson, S.D. (1975) The effects of psychomotor stimulants on stereotypy and locomotor activity in socially-deprived and control rats. *Brain Research* 84, 195–205.

Schlichting, M.C. and Smidt, D. (1984) The behaviour of gilts in relation to the time spent in different housing systems. In: Unshelm, J., van Putten, G. and Zeeb, K. (eds), *Proceedings of the International Congress on Applied Ethology in Farm Animals*. KTBL, Darmstadt, pp. 156–158.

Schwartz, B. (1980) The development of complex, stereotyped behavior in pigeons. *Journal of the Experimental Analysis of Behavior* 33, 153–166.

Schwartz, B. (1982) Failure to produce response variability with reinforcement. *Journal of the Experimental Analysis of Behavior* 37, 171–181.

Sroufe, L.A. and Cooper, R.G. (1988) *Child Development – Its Nature and Course*. Alfred A. Knopf, New York.

Staddon, J.E.R. and Ayres, S.L. (1975) Sequential and temporal properties of behavior induced by a schedule of periodic food delivery. *Behaviour* 54, 26–49.

Staddon, J.E.R. and Simmelhag, V.L. (1971) The 'superstition' experiment: a reexamination of its implications for the principles of adaptive behavior. *Psychological Review* 78, 3–43.

Stolba, A., Baker, N. and Wood-Gush, D.G.M. (1983) The characterisation of stereotyped behaviour in stalled sows by informational redundancy. *Behaviour* 87, 157–181.

Stone, A.A. (1964) Consciousness: altered levels in blind retarded children. *Psychosomatic Medicine* 26, 14–19.

Suomi, S.J. (1989) Genetic and environmental factors shaping individual differences in rhesus monkeys' behavioral development. *International Ethological Congress*, Utrecht, The Netherlands.

Stuss, D.T. and Benson, D.F. (1984) Neuropsychological studies of the frontal lobes. *Psychological Bulletin* 95, 3–28.

Terlouw, E.M.C., Lawrence, A.B. and Illius, A.W. (1991a) Influences of feeding level and physical restraint on the development of stereotypies in sows. *Animal Behaviour* 42, 981–992.

Terlouw, E.M.C., Lawrence, A.B., Ladewig, J., de Pasillé, A.M. and Schouten, W.G.P. (1991b) Relationship between plasma cortisol and stereotypic activities in pigs. *Behavioural Processes* 25, 133–153.

Terlouw, E.M.C., de Rosa, G., Lawrence, A.B. and Illius, A. (1993a) Behavioural responses to amphetamine and apomorphine in pigs. *Pharmacology, Biochemistry and Behavior* (in press).

Terlouw, E.M.C., Lawrence, A.B. and Illius, A. (1993b) Relationship between amphetamine and environmentally induced stereotypies in sows. *Pharmacology, Biochemistry and Behaviour* 43, 347–355.

Thelen, E. (1979) Rhythmical stereotypies in normal human infants. *Animal Behaviour* 27, 699–715.

van Putten, G. (1982) Discussion of Session III. In: Bessei, W. (ed.) *Disturbed Behaviour in Farm Animals*. Eugen Ulmer, Hohenheimer Arbeiten, Stuttgart, p. 129.

von Borell, E. and Hurnik, J.F. (1991) Stereotypic behaviour, adreno-cortico function and open field behaviour of individually-confined gestating sows. *Physiology and Behaviour* 49, 709–714.

Warburton, D.M. (1987) The neurobiology of stress response control. In: van Adrichem, P.W.M. and Wiepkema, P.R. (eds), *The Biology of Stress in Farm Animals: An Integrated Approach*. Martinus Nijhoff, Dordrecht, pp. 83–86.

Wechsler, B. (1991) Stereotypies in polar bears. *Zoo Biology* 10, 177–188.

White, P.J., Kreeger, T.J., Seal U.S. and Tester, J.R. (1991) Pathological responses of red foxes to capture in box traps. *Journal of Wildlife Management* 55, 75–80.

Wiepkema, P.R. (1984) Pain and stereotypies. In: Duncan, I.J.H. and Molony, V. (eds), *Agriculture: Assessing Pain in Farm Animals*, Proceedings of a Workshop held in Roslin, Scotland. CEC (Commission of the European Communities) pp. 62–70.

Wiepkema, P.R. (1987) Behavioural aspects of stress. In: van Adrichem, P.W.M. and Wiepkema. P.R. (eds), *The Biology of Stress in Farm Animals: An Integrated Approach*. Martinus Nijhoff, Dordrecht, pp. 113–134.

Wiepkema, P.R., Broom, D.M., Duncan, I.J.H. and van Putten, G. (1983) *Abnormal Behaviours in Farm Animals*. A report of the Commission of the European Communities, Brussels.

Wood-Gush, D.G.M. (1972) Strain differences in the response to sub-optimal stimuli in the fowl. *Animal Behaviour* 20, 72–76.

Wood-Gush, D.G.M., Stolba. A. and Miller, C. (1983) Exploration in farm animals and animal husbandry. In: Archer, J. and Birke, L.I.A. (eds), *Exploration in Animals and Humans*. Van Nostrand Reinhold, London, pp. 198–209.

The Motivational Basis of Stereotypies

JEFFREY RUSHEN[1], ALISTAIR B. LAWRENCE[2] AND E.M. CLAUDIA TERLOUW[2]

[1]*Agriculture Canada Research Station, Lennoxville, Quebec, Canada:* [2]*The Scottish Agricultural College, Edinburgh, UK.*

Editors' Introductory Notes:
Rushen *et al.* suggest that progress in understanding the behavioural basis of stereotypies can best be made, not by simply describing the behaviour as 'abnormal', but by acknowledging that processes involved in controlling normal behaviour may also underlie stereotypies. They review evidence that specific motivational states lie at the origin of stereotypies. The best evidence for this comes from work looking at the relationship between feeding problems and stereotypies which strongly implicates heightened feeding motivation as a strong causal factor in development of stereotypies in food-restricted animals. There is less evidence for a relationship between other motivational systems and stereotypies, although this most likely reflects a lack of research.

Although a specific motivation such as feeding may be implicated in development of stereotypies this tells us little about the exact nature of the relationship. Rushen *et al.* discuss the various motivational processes that might be involved in development of stereotypies. Some of these processes, such as negative and positive feedback, are related to specific motivations while others, such as arousal, may have more general (non-specific) effects across a number of motivations. Rushen *et al.* conclude that it is likely that a number of different motivational processes will be required to explain how activation of specific motivations results in stereotypic rather than normal behaviour and also the different properties of stereotypies.

At the present time, a reasonable hypothesis is that stereotypies are the expression of a motivational state modified or channelled by the environment. The restriction the environment imposes on behavioural expression may also be important in determining the degree of stereotypy. Persistence may result from positive feedback increasing underlying motivation in combination with an absence of sufficient negative feedback. Persistence could also arise from arousal reducing the tendency of animals to rest and increasing the tendency to engage in active behaviours.

Introduction

Stereotypic behaviours have had a high profile in attempts to develop behavioural indicators of animal welfare and they are often described as 'abnormal'. However, such a description does little to further understanding. In this chapter, we suggest that the motivational processes that govern 'normal' behaviour can also be used to understand stereotypic behaviour (see also Mason, 1991). There is evidence (Fentress, 1973, 1976) that under some conditions motivational systems have an inherent tendency to become 'self-organized'. Self-organized systems are those that are less easily disrupted by environmental input and which involve highly predictable sequences of behaviour. For example, sequences of grooming by mice contain rapid phases which are more stereotyped and less easily disrupted than the slower and more complex movements that are found at the beginning and end of the sequence (Fentress, 1973). Since these are some of the properties that define abnormal stereotypic behaviour (Ödberg, 1978; Chapter 2), it is tempting to suggest that such normal processes can explain the occurrence of abnormal stereotypies (Fentress, 1976).

The motivational analysis of behaviour is the study of those causal factors that underlie the expression of behaviour. These causal factors can be broadly divided into those that relate to specific motivational states (e.g. to feeding) and other, less specific factors (e.g. arousal) that can modulate the expression of a number of different motivational states. The emphasis in studies of motivation is often on the short-term control of behaviour, but longer term factors, such as those involved in learning, clearly have influences on motivation (Bindra, 1976). Motivational analysis is a 'black box' approach to understanding behaviour, involving the development of hypothetical systems or models representing the unobservable controls on behaviour.

A number of motivational models have been developed to explain the observed behavioural properties of stereotypic behaviour. We review the behavioural data on stereotypies and examine the success of these motivational models. Despite the interest in stereotypies, the number of behavioural studies is quite low and the data have tended to come from descriptive rather than experimental research (Chapter 1). Both factors may explain the relatively slow rate of advance in our understanding of the motivational bases of these behaviours. Furthermore, existing explanatory models of stereotypies have tended to rely on a small number of motivational processes to explain stereotypies in general. In contrast, we suggest that a number of different motivational processes are required to explain different aspects of stereotypic behaviour.

Specific Motivational Systems Underlying Stereotypic Behaviour

Stereotypic behaviour is found most often where animals are confined and where their behaviour is restricted (Kiley-Worthington, 1977; Mason, 1991). This has led to the suggestion that frustration of specific motivational systems, or the resulting aversion, may underlie stereotypies. In this section, we will review the empirical evidence that specific motivational states underlie stereotypies.

FEEDING

There is a growing number of examples of a relationship between stereotypies and feeding and foraging behaviour. This is particularly true for stereotypies associated with feeding periods (Rushen, 1985; Terlouw et al., 1991a, b).

At the simplest level, the occurrence of stereotypies may reflect inadequate nutrient intake (see Lawrence et al., 1993). Food restriction has been found to increase the incidence of stereotypies in pigs (Appleby and Lawrence, 1987; Appleby et al., 1989; Terlouw et al., 1991a), sheep (Marsden and Wood-Gush, 1986) and poultry (Savory et al., 1992). Food restriction and its behavioural consequences are important causal factors even in stereotypies that occur after feeding (e.g. Rushen, 1984). For example, pregnant sows are routinely food-restricted to approximately 60% of their *ad libitum* intake and this level of feeding has been shown to result in high levels of feeding motivation throughout the day, even in the immediate post-feeding period (Lawrence et al., 1988; Lawrence and Illius, 1991) perhaps reflecting positive feedback from the ingestion of food (de Passillé et al., 1992). The effect of food restriction on feeding motivation probably also increases in the long term, perhaps as a result of an accumulating nutrient deficit (see Lawrence et al., 1988; Terlouw et al., 1991a).

Feeding motivation is affected by a large number of pre- and post-absorptive factors (Le Magnen, 1985) and it is important to understand which of these is responsible for the increased hunger that is now implicated in stereotypies. For example, Rushen (1984) suggested that a lack of food bulk might be important in the development of stereotypies. Concentrated food increases the incidence of stereotypic behaviour in ruminants (Willard et al., 1977; Kooijman et al., 1991). Increasing food bulk with sugar beet has been found to decrease stereotypies in pigs (Brouns et al., 1991), while ground straw has no effect (Fraser, 1975). Oat hulls have been found to reduce (Robert et al., 1992) or to have no effect on stereotypies in pigs (Broom and Potter, 1984). These differences may reflect the capacity of different fibres to decrease feeding motivation: for example, adding ground straw to pigs' diets does not reduce food intake (Brouns et al., 1991) or operant responding for food (Lawrence et al., 1989), whereas addition of sugar beet does reduce intake (Brouns et al., 1991). The relative importance of specific nutrient deficits on hunger and the development of stereotypies has received little attention, although specific calcium deficiency has been implicated in stereotypic air-pecking in hens (Hughes and Wood-Gush, 1973).

Although food restriction and hunger are related to the development of stereotypies, this tells us little about the exact relationship. For example, stereotypies could arise from the increased appetitive foraging behaviour that accompanies food restriction (cf. Hughes and Duncan, 1988); increasing the opportunity for foraging reduces stereotypies in canaries (Keiper, 1969), bears (Carlstead et al., 1991), walruses (Kastelein and Wiepkema, 1989), and primates (Line et al., 1989a; Lam et al., 1991).

Clearly, feeding problems can play a role in the occurrence of some forms of stereotypic behaviour. However, to understand the exact nature of the relationship, we need to have a clearer understanding of the complex nature of feeding motivation in different species.

Locomotory behaviour

Morris (1964) and Hediger (1955) suggested that the stereotyped pacing of zoo animals might be derived from normal patrolling of the territory, which becomes stereotyped because of the limited space available. However, locomotion can arise from a number of motivational states. For example, deprivation of both food and dust-bathing increased exploratory locomotion in poultry (Nicol and Guilford, 1991). Further, as discussed below, aversive environments can give rise to locomotory escape attempts. The effect of increasing space allowance on stereotypies should therefore depend on the motivational basis of the locomotion and so increased space allowance need not necessarily reduce stereotypies. In support of this, manipulating space allowance has been found to have variable effects on stereotypies. Increased space allowance has been reported to reduce route-tracing by canaries (Keiper, 1969), stereotypic behaviour in some primates (Draper and Bernstein, 1963) and horses (Krzack et al., 1991), and sham-chewing in pigs (Terlouw et al., 1991a). However, space allowance was found to have less effect than environmental enrichment on stereotypies in voles (Ödberg, 1987a), and chain-chewing and excessive drinking by pigs was not affected by space allowance (Terlouw et al., 1991a). Some studies have found a lack of affect of increased space allowance on stereotypies in primates (Line et al., 1989b, 1990). Therefore, while some findings are consistent with the idea that stereotypies develop when locomotion is restricted, it is clear that this interpretation cannot be generalized to all forms of stereotypies.

Aversion

Many forms of stereotypies, especially pacing or locomotory stereotypies, appear to be derived from attempts by the animal to escape from its environment (Morris, 1964; Meyer-Holzapfel, 1968; Duncan and Wood-Gush, 1972a). In one of the few experimental tests of this idea, Duncan and Wood-Gush (1972a) found that physically preventing hens from reaching food that they could see resulted initially in escape attempts and then later in stereotypic pacing. They concluded that the stereotypic pacing was derived from the attempts to escape from an aversive environment.

In his influential thesis, Cronin (1985) argued that stereotypies of tethered sows develop from escape attempts and aggressive behaviour elicited by the aversion and frustration of physical restraint. This interpretation was based on his description of the stereotypies developing from the initial reactions to tethering, which forms the most thorough developmental study yet undertaken. The interpretation was supported by his finding that the original aggressive behaviour reappeared when stereotypies were blocked by the opioid antagonist, naloxone. However, there are a number of problems with Cronin's interpretation of his observations. First, Terlouw *et al.* (1991a) have shown convincingly that tethering is not a prerequisite for stereotypies to develop in sows. Second, stereotypic behaviour in sows involves oral activities such as chewing, rooting, sham-sucking and drinking, and it is not clear from Cronin's account how these behaviours are derived from escape behaviours such as pulling on the tether, screaming, and violent head movements. Third, other developmental comparisons have resulted in different conclusions: using an age-comparison, Rushen (1984, 1985) concluded that stereotypies result from feeding behaviour, while Stolba *et al.* (1983) concluded that exploratory behaviour was the source. Clearly, there is a considerable amount of judgement involved in deducing underlying motivation from the appearance of the behaviour. Fourth, Schouten and Rushen (1992) found that when stereotypies were inhibited by naloxone, rooting behaviour occurred rather than aggressive behaviour. Finally, the individuals that react most to initial tethering are least likely to develop stereotypic behaviour (Schouten and Wiepkema, 1991).

Clearly, although many feel that stereotypies are a reaction to aversive environments, considerable care is required before assuming that aversion always underlies stereotypies or that all stereotypies are derived from escape attempts. There is a need for more experimental analysis to demonstrate that the environments really are aversive and that manipulation of the degree of aversiveness affects the occurrence of stereotypies.

The relationship between the occurrence of stereotypies and the aversiveness of the environment might not, however, be straightforward. Cooper and Nicol (1991) have shown that, when voles are given the choice between an enriched environment and a barren and possibly aversive environment, voles with little stereotypic behaviour show a preference for the enriched environment. Voles showing a high level of stereotypic behaviour, however, express much less of a preference. The reason for this difference is unclear; Cooper and Nicol (1991) suggest that the lack of preference is due to the stereotypies reducing the animals' perception of aversion. This interesting observation should stimulate further enquiry. However, it may be that the enriched environment impedes the performance of stereotypies and that in order to perform stereotypies the voles must move to the barren environment. Alternatively, voles that develop stereotypies may find both environments aversive and so show little preference (Rushen, 1993).

OTHER MOTIVATIONAL STATES

Stereotypies can be reduced by various forms of environmental enrichment (e.g. Ödberg, 1987a; Bryant *et al.*, 1988; Lam *et al.*, 1991) that do not seem closely related to aversion, feeding or locomotory behaviour. This suggests environmental restriction of other behaviours besides those mentioned. Carlstead and Seidensticker (1991) suggested that some forms of stereotypies by bears were related to sexual motivation since the behaviours tended to occur during seasons of the year when mating normally occurred and were reduced somewhat in male bears by the presence of female odours. In laying hens and farrowing sows, nest-building motivation is thought to underlie stereotypies, particularly those that occur around oviposition and parturition (Wood-Gush, 1972; Baxter, 1982; Jensen, 1988). Wood-Gush and Vestergaard (1989) and Stolba *et al.* (1983) claimed that stereotypies in pigs are related to exploratory motivation. However, in many cases these are simply suggestions which are not based on sufficiently robust experimental evidence. Furthermore, attempts are rarely made to exclude alternative explanations. An exception is the work of Lam *et al.* (1991) who showed that increased opportunities for grooming reduced stereotypic behaviour in macaques.

CONCLUSION

There is growing evidence that specific motivational states that control normal behaviour play a role in the development of stereotypies. It seems most likely that different forms of stereotypic behaviour will be affected by different motivational states. However, there is a lack of detailed evidence implicating motivational states other than those controlling feeding. Further, we have yet to consider why activation of motivational states results in stereotypies rather than normal behaviour. For example, Terlouw *et al.*

1993a) have argued that the most appropriate response of hungry pigs to the absence of food is to rest and feral pigs have been observed to increase resting time and decrease foraging time during food scarcity (Graves, 1984). Clearly, the high levels of activity and stereotypic behaviour seen when tethered sows are food-restricted (Rushen, 1985; Cronin *et al.*, 1986; Appleby and Lawrence, 1987; Terlouw *et al.*, 1991b) cannot be due solely to the activation of the feeding motivational system (Hughes and Duncan, 1988). To understand how activation of motivational states can result in stereotypic behaviour, we need to consider in more detail the various motivational processes controlling behavioural expression.

Motivational Processes of Relevance to Stereotypies

INTERNAL SOURCES OF MOTIVATION

The motivational states that underlie behaviour receive input from external and internal sources. Behavioural deprivation has been a central issue in animal welfare (Dawkins, 1983, 1988, 1990; Hughes and Duncan, 1988). Consequently, applied ethologists have focused their attention on internal sources of motivation. The term 'behavioural need' is often used implying a strong internally driven motivation to perform behaviour (Dawkins, 1983) and which might indicate frustration if that behaviour is prevented by the environment in which the animal is kept (e.g. Nicol, 1987; Hughes *et al.*, 1989; Wood-Gush and Vestergaard, 1989).

Lorenz (1981) produced a particularly influential motivational model which described behaviour as motivated by a combination of internal and external factors. The uniqueness of his model was that it suggested that animals are motivated to perform particular behaviour patterns and not just to achieve particular functional goals. For example, an animal that has consumed sufficient food to meet nutritional requirements may still need to perform appetitive food-searching behaviours. Furthermore, Lorenz suggested that the tendency to perform these behaviour patterns increases with the time since their last performance as a result of an accumulation of internal 'nervous energy'. At normal levels of motivation, the behaviour will be elicited only by stimuli that are close to the optimal. As time passes and the internal motivation rises, the behaviour will begin to be elicited by a wider range of suboptimal stimuli. Eventually, the motivation is so high that the animal performs the behaviour in the absence of any eliciting stimuli.

The Lorenzian model of motivation has been attacked many times, largely due to the postulated source of nervous energy, which conflicts with what is known of the physiology of the nervous system (Hinde, 1970; McFarland, 1971). Applied ethologists have been heavily criticized for continuing to accept such an apparently outdated model (Dawkins, 1983). Despite this criticism, Lorenz's ideas continue to have some appeal in applied ethology, largely because 'sham' or 'vacuum' activities are often seen in confined animals (e.g. Black and Hughes, 1974; Vestergaard, 1982; Van Liere, 1991). Recently, Lorenz's ideas have been sympathetically re-evaluated. Toates (1986) and Toates and Jensen (1991) argue that it is possible to dismiss the dubious 'energy' mechanism while accepting that certain parts of Lorenz's model do correspond with aspects of animal behaviour. For example, there is evidence that the tendency to perform a certain behaviour does increase with time since last performance (i.e. they show 'rebound') (Nicol, 1987). Furthermore, in some cases animals do seem to need to perform particularly behaviour patterns rather than just attaining the goals that those behaviour patterns usually achieve (Hughes and Duncan, 1988). For example, a hen provided with a replica of a nest she has just built continues to perform many of the behavioural patterns involved in nest-building, even if these do not alter the nest in any way (Hughes et al., 1989). Clearly, there are aspects of behaviour related to Lorenz's model that require explanation, and these appear to have great relevance to the study of stereotypic behaviour. A build-up of internal sources of 'action-specific energy' has often been postulated to underlie stereotypic behaviour, which is considered a 'sham' activity (e.g. Sambraus, 1985).

However, considerable caution should be exercised before accepting Lorenz's model, even in part. First, the model need not apply to all motivational systems and its applicability must be decided on a case by case basis. Second, it remains difficult to exclude the possibility that the rebound in behaviour following a period of deprivation is not a response to the novelty of the altered environment rather than to an increase in internal sources of motivation (McFarland, 1989). Third, subtle and unexpected external cues may be evoking the apparent 'vacuum' behaviour (Van Liere, 1991) and many stereotypies appear to be elicited by an external event such as feeding (e.g. Rushen, 1984). Fourth, there is little to be gained in simply postulating internal 'Lorenzian' sources of motivation to explain stereotypic behaviour, without more evidence that the motivation to perform the behaviour does increase with time, and that only the performance of the behaviour serves to reduce the motivation. Many types of stereotypies only develop in food-restricted animals (Appleby and Lawrence, 1987; Terlouw et al., 1991a: Savory et al., 1992), suggesting that the tendency to perform the behaviour is at least partly related to a metabolic deficit and does not increase in the absence of this deficit.

In conclusion, decades of research into motivation has shown that internal and external causes are closely interwoven and it is unlikely that they can be clearly separated (Toates, 1986). Even for invertebrates such as *Aplysia*, the internal state alone is not sufficient to maintain feeding behaviour which requires some external eliciting stimulus (Kandel, 1979). Furthermore, changes in internal state can cause changes in responsiveness to external cues (Hinde, 1970). It is not useful therefore to think of independent internal or external controls of behaviour (Toates, 1986), although it may be that the balance of internal and external control is greater for some motivational states than others (Hughes and Duncan, 1988; Toates and Jensen, 1991). Any adequate model of stereotypies must consider the influence of both internal and external causes.

INFLEXIBILITY OF MOTIVATIONAL SYSTEMS

There are a number of examples that suggest the sequences of behaviour used by animals to achieve functional goals are not completely flexible. Breland and Breland (1961) describe how pigs, trained to pick up a token and deposit it in a box in order to obtain food, gradually came to spend more and more time rooting at the token, to the extent that they began to miss out on food rewards. They termed this 'misbehaviour'. Domestic fowl when waiting for food to appear began to show distinctive ground-scratching, even if this did nothing to hasten the arrival of food. It has often been observed in studies of operant conditioning that animals will 'auto-shape', that is the animals often direct appetitive or consummatory behaviour to particular parts of the operant chamber even if this is not required for the delivery of food (Brown and Jenkins, 1968). Breland and Breland (1961) found that increasing the underlying motivational state increased the tendency of the animals to perform these non-functional appetitive sequences (see also Timberlake, 1983).

There is evidence for some degree of autonomy of appetitive behaviour and its uncoupling from consummatory aspects. Morgan (1974) has described a number of examples where appetitive feeding responses continue in satiated animals, suggesting that appetitive behaviours show some resistance to satiation. These examples indicate that it is difficult to suppress the appetitive sequences of behaviour associated with a motivational state, particularly as the strength of the underlying motivation increases.

Hughes and Duncan (1988) argue that many stereotypies involve appetitive sequences which the animals must perform even if their performance is unnecessary to achieve functional goals. Hughes and Duncan (1988) suggest that activities such as appetitive foraging can be positively reinforcing and increase the strength of the underlying motivational state. Where there is insufficient negative feedback from the functional consequences of the consummatory feeding behaviour, these appetitive sequences will persist, developing into stereotyped behaviour. There is some evidence to support this view, Morris (1964) suggested that animals need to be able to perform normal foraging even if not food-restricted. As mentioned above (p. 44), there is evidence that increasing the opportunity for foraging reduces stereotypic behaviour.

Interestingly, there are species differences in the timing of stereotypies relative to the feeding periods. While carnivores tend to perform stereotypies before feeding (Meyer-Holzapfel, 1968), other species such as poultry (Kostal et al., 1992), pigs (Rushen, 1985; Terlouw et al., 1991a, b) and pigeons (Palya and Zacny, 1980) tend to show most stereotypies after feeding. This may reflect species differences in the normal temporal organization of appetitive and consummatory behaviour (Terlouw et al., 1991a).

However, in many cases, animals can be very flexible in how they vary their behaviour according to the context (e.g. Timberlake, 1983). In order for this explanation of stereotypies to be useful, it must be shown that stereotypies tend to occur in situations where normal behaviour tends to be most inflexible. There are differences between the circumstances in which 'misbehaviour' of animals is observed and the conditions in which stereotypies develop. For example, 'misbehaviour' occurs where animals continue to receive some rewards, albeit at a low rate. Although 'misbehaviour' can lead to the animal delaying the appearance of rewards (Breland and Breland, 1961; Timberlake, 1983), there are no reports of its persistence in the long-term absence of rewards (Gardner and Gardner, 1988), as is the case for stereotypic behaviour. Furthermore, we cannot assume that stereotypies are always derived from appetitive foraging sequences. For example, Rushen (1984) and Schouten and Rushen (1992) suggested that post-feeding stereotypies in pigs were derived from rooting. However, Terlouw et al. (1991a) found in pigs that while levels of stereotypic behaviour increased over time, appetitive rooting and nosing at the trough was not replaced, suggesting that this behaviour was not the source of the stereotypic behaviour.

ENVIRONMENTAL 'CHANNELLING' OF BEHAVIOUR

It seems likely that the physical environment will strongly influence the flexibility of motivational systems by restricting the sorts of behaviours that the animal can express. Lawrence and Terlouw (1993) have suggested that, in a strongly motivated animal, behaviour will be modified or 'channelled' by the environmental restrictions. The resulting behaviour will therefore reflect the limitations that the environment places on behavioural expression as well as the underlying motivational state. Lawrence and Terlouw (1993) also suggest that this environmental channelling may help explain one of the most striking qualities of stereotyped behaviour, that it is often very fixed in form. Rigid and inflexible behaviour may result in part from static and simple environments. As an example, we know that under relatively unrestricted conditions, such as at the Edinburgh Pig Park, foraging by pigs involves a variable and complex set of behaviours. Under the more restricted conditions of a stall, this behaviour can only be expressed against a few static substrates such as chains and bars. Potentially complex behaviour is therefore channelled and simplified by the unvarying characteristics of the environment. A similar point has been made by Hediger (1955) and Morris (1964) who point out that the pacing of large carnivores in a small cage cannot be expected to be particularly variable due to the restrictions placed on the behaviour by the environment.

A prediction of this hypothesis is that restricting the expression of different motivational states will result in similar stereotypies in the same environment, and that restricting expression of the same motivational state will produce different stereotypies in different environments. A similar idea has also been suggested by Dantzer (1986). However, careful observations have revealed that the degree of stereotypy varies considerably over time and between animals in the same environment (Stolba et al., 1983; Cronin and Wiepkema, 1984). Consequently, channelling alone will not explain the rigidity of stereotypic behaviour.

TRANSITIONS BETWEEN MOTIVATIONAL SYSTEMS

The behaviour of an animal does not reflect the strength of a single motivational system as, at any one time, the motivation for performing many incompatible behaviours will be present. For example, a hungry animal that has just discovered food may suddenly become aware of an approaching predator. The animal has to make a decision as to which behaviour has the highest priority. The manner in which animals make decisions about which behaviour to perform is one of the most difficult areas in motivational research and one that has received much attention in recent years. According to McFarland and his co-workers (e.g. McFarland and Houston, 1981), the choice between different behavioural options reflects the relative strengths of the different motivational systems that are competing for expression. As one motivational goal is satisfied the animal will switch its behaviour to the next highest priority behaviour. Consequently, the interaction between different motivational systems will be apparent in the way that animals switch from one behaviour to another (McFarland, 1976), and will determine how much time the animal spends in each activity. McFarland (1969) distinguished between two types of causal processes that result in switching between activities. In the first process, 'competition', switching results from a gradual increase in the underlying motivation for a second activity which inhibits the on-going activity. The second process, 'disinhibition', involves a decrease in the motivation underlying the current activity, which then allows expression of the second activity.

It is important to note, therefore, that the amount of time animals spend in stereotypic behaviour will depend not just on the strength of the underlying motivational system but also upon the relative strengths of competing motivational systems. This point is often ignored in discussions of stereotypic behaviour, although Hughes and Duncan (1988) are an exception. These authors have raised the interesting speculation that stereotypies may result because many captive or domestic animals do not have sufficient opportunities for performing behaviours that might compete with stereotypic behaviour for the animal's time. In the rather barren environments in which captive animals are held, environmental changes are infrequent, and the lack of stimuli eliciting competing behaviours may then be partly responsible for stereotypic behaviour (Kiley-Worthington, 1977). Fraser (1975) suggested that straw given to pigs reduces the duration of stereotypic behaviour by a 'recreational effect': that is the straw acts by increasing the time spent in alternative activities.

There is, in fact, a variety of evidence that sudden changes in an animal's environment can disrupt stereotypic behaviour (Wemelsfelder, 1990), which is often taken as evidence that stereotypies arise in response to boredom. The concept of boredom is intuitively attractive as an explanation for stereotypies, as the conversation of the public at zoo exhibits can reveal. Boredom has however lacked sufficiently well-defined criteria, which has limited its scientific use (but see Wemelsfelder, 1990, and Chapter 4). Furthermore, if simple boredom or lack of stimulation were responsible for stereotypies, one would not expect the dramatic declines in stereotypies that result when the strength of specific motivational systems are selectively reduced (e.g. Appleby and Lawrence, 1987; Terlouw *et al.*, 1991a).

NEGATIVE AND POSITIVE FEEDBACK

It has become increasingly common to model motivational processes on the principles of control systems (McFarland, 1985; Toates, 1986). An essential aspect of control system models is the use of negative feedback loops that link underlying motivational tendency to the functional consequences of the behaviour. Once a behaviour, such as feeding, is initiated it continues until monitored feedback from the consequences of the behaviour (e.g. food intake in the case of feeding behaviour) reduces the underlying motivational tendency and switches off the behaviour.

However, there are problems with models of motivation that rely solely on such negative feedback. As discussed in the previous section motivational models based on control systems generally assume that the behaviour currently being performed (behaviour 'A') has the highest motivational tendency of all the alternative behaviours. Behaviour 'A' is replaced when negative feedback causes its tendency to fall below that of the activity (behaviour 'B') with the next highest tendency. When the tendency of behaviour 'A' falls just below the tendency of the alternative behaviour 'B', the animal should switch to behaviour 'B'. However, since the tendency for 'B' is only slightly above that of 'A', the animal will only have to engage in this behaviour for a short time before its tendency is again below that for 'A'. The resulting oscillation or 'dithering' between activities is repeatedly found in simulations of behaviour that rely only on negative feedback (Ludlow, 1980), but it is rarely observed in real animals. One possible reason for the absence of dithering is the existence of positive feedback (McFarland, 1971) where the tendency to perform a behaviour actually increases as a result of its performance. There is some evidence for the existence of positive feedback in feeding and drinking (Wiepkema, 1971; Houston and Sumida, 1985) and suckling (de Passillé *et al.*, 1992).

Positive and negative feedback may be particularly important in explaining the prolonged duration of stereotypic behaviour, which is another striking quality of this behaviour. Cronin *et al.* (1986) reported sows spending an average of 8 hours a day in stereotypies; Rushen (1984) saw continuous bouts of stereotypic behaviour lasting as long as 2 hours; Ödberg (1986) reports bank voles showing stereotyped jumping several thousand times per day. Hughes and Duncan (1988) suggest that stereotypic behaviours are derived from sequences of appetitive behaviour, which continue partly as a result of positive feedback from the performance of the behaviour and partly because the animal is unable to proceed to the consummatory phase and so receive negative feedback. Wiepkema (1987, 1990), again using a control systems approach, suggested that behaviour is motivated by a discrepancy between the optimal or set-point value of certain organismic and environmental variables ('Sollwert') and their actual value ('Istwert'). Each species has a limited range of behaviours used to reduce this discrepancy. If these attempts do not succeed the behaviours continue to be performed anyway and eventually develop into stereotypic behaviour (see also Ödberg, 1987b).

However, clear examples where stereotypic behaviour has resulted from positive feedback or from the absence of negative feedback are few. De Passillé *et al.* (1992) showed that object sucking by calves increased after ingestion of milk. Duncan and Wood-Gush (1972a) showed that the inability to consume food, when feeding was elicited by the sight of food, resulted in stereotypic pacing by hens; Lawrence *et al.* (1988) showed that the amount of food eaten by sows was insufficient to reduce their feeding motivation, which could be responsible for post-feeding stereotypic behaviour in sows. More research is needed into the role of negative and positive feedback in affecting the duration of stereotypic behaviour.

CONFLICT AND DISINHIBITION

Sometimes an optimal sequencing of competing behaviours cannot be achieved, and the resulting motivational conflict disrupts the animal's behaviour. 'Displacement' behaviours often occur when two or more strongly motivated behaviours are in conflict, or when an animal is physically prevented from performing a strongly motivated behaviour. For example, if a stuffed owl is placed near to some chaffinches in a cage, the chaffinches are motivated both to flee from and to approach and mob the owl. The approach–avoidance conflict is apparent in the unusual amount of seemingly irrelevant preening they perform (Rowell, 1961). There are a number of similarities between these displacement activities and stereotypic behaviour. First, displacement behaviours have no obvious function and are often described as occurring in stereotyped sequences (e.g. Bindra, 1959). They are often performed in a bizarre and exaggerated manner, and at a high rate (Duncan and Wood-Gush, 1972b; Roper, 1984). It has often been suggested that stereotypic behaviour is a form of displacement behaviour reflecting motivational conflict in animals (e.g. Duncan and Wood-Gush, 1972b; Wiepkema, 1985; Ödberg, 1987b).

However, characterizing stereotypies as displacement behaviours is not helpful unless we understand why displacement activities occur. The 'disinhibition' hypothesis of displacement behaviour suggests that the motivational conflict reduces the tendency to engage in the two conflicting behaviours, so that the tendency to perform a third behaviour (the displacement behaviour) becomes relatively stronger (Andrew, 1956; McFarland, 1966; Pring-Hill, 1979). More specifically, McFarland (1966) suggests that the animal switches attention away from cues relating to the blocked behaviours to cues relevant to the displacement behaviour. A prediction of this is that the displacement behaviour will be more likely to be affected by cues related to its own motivational system than by cues related to the blocked motivational systems. There is some evidence for this (McFarland, 1965). However, the displacement hypothesis cannot always explain which particular behaviours will occur as displacement activities. Fentress (1973) suggests that grooming behaviours often occur as displacement behaviours because they require less environmental control, and can therefore more easily be performed out of context. The disinhibition theory also cannot explain the enhanced rate at which the behaviours are performed (Duncan and Wood-Gush, 1972b; Roper, 1984). These problems also apply to the use of the disinhibition theory to explain stereotyped behaviour.

Arousal

Arousal has often been invoked to explain properties of stereotypies. For example, Ödberg (1987b) suggested that arousal arising from behavioural frustration was a common factor that linked different forms of stereotypies. Furthermore, the concept of arousal has often been used to link behavioural aspects of stereotypies with physiological measures (Chapters 6 and 7). In a comprehensive review of the role of arousal in stereotypic behaviour, Dantzer (1986) suggested that the general arousal engendered by the environment elicits behavioural responses that depend on the environmental stimulation and disposition of the animal. The particular behaviours incorporated into stereotypies may simply be the behaviours that happen to be occurring when the animal is subjected to some arousal. Fentress (1976) suggested that self-organization of many behavioural systems increases with an increase in the level of arousal.

In the behavioural sciences, the concept of arousal has had a chequered career. One problem is that different writers mean different things by the term. Arousal was first linked to cerebral activation, specifically to low-voltage, fast electrical activity in the neocortex and rhythmical slow activity in the hippocampus (Lindsley, 1960). However, a straightforward relationship between arousal and brain activity is no longer accepted (e.g. Vanderwolf and

Robinson, 1981) and in the eyes of some 'the old arousal theory' has finally died. Some use the term simply to indicate that an animal is not asleep. Some use it to refer to processes leading to an increased rate of performance of behaviour (Killeen *et al.*, 1978). Others relate arousal to endocrine systems such as pituitary–adrenocortical activity (Hennessy and Levine, 1979). Unfortunately, different behavioural and physiological measures of arousal often do not correlate; even in *Aplysia*, the selective effects of appetitive and noxious stimuli indicate that arousal is not a unitary process (Kandel, 1979). With all of its accumulation of meanings, use of the term arousal tends to confuse rather than enlighten.

Despite problems with the concept of arousal, there are examples from mammals of non-specific motivational processes which appear to activate a number of behavioural systems (Roper, 1980). For example, when animals are mildly disturbed they engage in many consummatory behaviours that are unrelated to the nature of the disturbance. Rats removed from their home cage and handled for a few minutes often begin eating when returned (Roper, 1980). When drinking behaviour is thwarted, rats engage in a number of activities such as eating, grooming and exploration, all of which are performed at a faster rate than normal (Roper, 1984). Mild stimulation, tail-pinch, electrical stimulation of the hypothalamus, and intermittent delivery of food can all increase the occurrence of many apparently unrelated types of behaviour (Roper, 1980). We use the term 'behavioural arousal' to refer to these non-specific activating factors.

One test of the role of arousal in stereotypies would be the ability of a range of different stimuli to elicit the same stereotypies. Fentress (1976) described how a range of stimuli could increase the performance of locomotory stereotypies in zoo animals. However, Terlouw *et al.* (1993b) found that stereotypies in pigs were not elicited by loud novel sounds, but only by the ingestion of food. Another contribution of arousal to stereotypies could be in reducing the tendency of the animals to rest (Terlouw *et al.*, 1993b). Brief feeding episodes have been shown to increase general activity in food-restricted animals (Killeen *et al.*, 1978). There is a certain amount of circumstantial evidence to link stereotypies with arousal in this way. For example, among food-restricted sows, the individuals that perform stereotypies tend to show higher levels of activity (Rushen, 1984; Von Borell and Hurnik, 1991). A role of arousal in stereotypies cannot be excluded, but apart from physiological studies (see Chapters 6 and 7) there have been too few direct tests. Moreover, non-specific motivational effects are probably too complex to be explained by single concepts like arousal and may still require to be explained by models of the interaction of specific motivational systems (Hinde, 1970).

INDIVIDUAL DIFFERENCES

There are substantial individual differences in the performance of many behaviours, and stereotypies are no exception. A number of studies have reported substantial variation between individuals in the level (e.g. Ödberg, 1986; Appleby and Lawrence, 1987) and type (e.g. Cronin, 1985) of stereotypic behaviour performed. Individual differences have traditionally been viewed as something of a statistical problem. However, more recently, there has been a growing interest in the use of individual differences to investigate the processes underlying behaviour, for example by investigating the physiological and neurological differences between individuals showing different behaviour. Critical to the study of individual differences is the demonstration that the same individual shows a consistency of behavioural response in different contexts at the same age (e.g. Lawrence et al., 1991) and across different ages (e.g. Kerr and Wood-Gush, 1987).

It has recently been shown that individual mice and rats differ in their tendency to show self-organized behaviour. A series of experiments have shown that certain individuals tend to be less responsive to the characteristics of their opponents during aggressive interactions, to show little response to environmental changes during the performance of learned behaviour and to show a more strongly entrained circadian rhythm (Benus, 1988). These differences may be related to the propensity of a certain individual to develop sterotypic behaviour. For example, Terlouw et al. (1991b) found in sows that aggressive tendencies measured in a food competition test predicted the development of excessive drinking, but not of chain-manipulation, during a subsequent period of housing and food restriction. At a neurological level, it has been proposed that the tendency to show self-organized behaviour reflects individual differences in sensitivity of the dopaminergic system. In pigs, behavioural responses to the indirect dopamine agonist, amphetamine, have been shown to be related to the subsequent development of both excessive drinking and chain-manipulation under restricted feeding and housing conditions (Terlouw et al., 1993b). These early results suggest that it may be possible to identify those motivational and neurological factors that account for individual differences in stereotypies, and in so doing help distinguish behavioural and neural mechanisms underlying stereotypies. Many have suggested that the search for individual differences that predispose an animal to develop stereotypies will be a major part of future research (e.g. Schouten and Wiepkema, 1991; Chapter 7).

Conclusions

There is considerable evidence that specific motivational states lie at the very origin of stereotypic behaviour. This is particularly true for feeding motivation, where both inadequate nutrient intake and inadequate opportunity for foraging can increase the incidence of stereotypies. There is less evidence for the involvement of aversion, locomotory or other motivational systems, although in most cases this reflects a lack of research, and different types of stereotypic behaviour are likely to arise from different causes. Further, a number of different motivational processes might be required to explain why activation of motivational systems results in stereotypic rather than normal behaviour. To date, no single motivational model has considered the full range of motivational processes that could contribute to stereotypies. It seems likely that different motivational processes will also be required to explain the different properties of stereotypic behaviour.

ORIGINS OF STEREOTYPIES

A number of examples have shown that stereotypies tend to arise when the normal behavioural expression of a motivational state is restricted or blocked. Stereotypies may then be appetitive or consummatory behaviour occurring in an unusual context. In this case, the nature of the motivational state will explain which behaviours are incorporated into stereotypies. However, there are a number of reasons why it is not always safe to infer the cause of the problem from the type of behaviour being shown. First, appetitive behaviours are often common to a number of different motivational systems and so there is considerable judgement needed to decide what motivational system is operating. Second, stereotypic behaviour could also arise from the disinhibition of secondary motivational states. Further, the behavioural content of stereotypies will reflect the interaction between the underlying motivational state and the environmental modification or channelling of that state. Channelling could affect the expression of both redirected and disinhibited behaviour. In highly restrictive environments the final behavioural expression may bear little resemblance to the expression of the state under less restrictive conditions.

DEGREE OF STEREOTYPY

The degree of stereotypy may be largely determined by the extent to which environments simplify and reduce the variability of behavioural expression. However, there is evidence that motivational systems have an inherent tendency to become self-organized particularly when the underlying motivational strength is high or when the animal is aroused. Further, individual animals differ in the extent that their behaviour becomes self-organized and independent of environmental cues.

PERSISTENCE OF STEREOTYPIES

The repetitive nature of stereotypies may partly reflect processes involved in specific motivational states. Positive feedback from the performance of the behaviour could increase the underlying motivation, and in the absence of sufficient negative feedback, this could lead to a persistently heightened motivational state. However, persistence could also result from arousal reducing the tendency of animals to rest and increasing the tendency to engage in active behaviours.

Unfortunately, much of this chapter has been speculative and a major problem has been the lack of detailed, fundamental research exploring the motivational factors underlying stereotypies. Motivational processes, particularly 'Lorenzian' type processes, are often evoked to explain stereotypic behaviour with inadequate experimental analysis to determine if they really are applicable. Applied ethologists often fail to appreciate the difficulties with such concepts. Until these problems are overcome, we do not foresee any rapid advance in the understanding of the motivational bases of these behaviours.

References

Andrew, R.J. (1956) Some remarks on behaviour in conflict situations with special reference to *Emberiza* sp. *British Journal of Animal Behaviour* 4, 41–45.

Appleby, M.C. and Lawrence, A.B. (1987) Food restriction as a cause of stereotypic behaviour in tethered gilts. *Animal Production* 45, 103–110.

Appleby, M.C., Lawrence, A.B. and Illius, A.W. (1989) Influence of neighbours on stereotypic behaviour of tethered sows. *Applied Animal Behaviour Science* 24, 137–146.

Baxter, M.R. (1982) The nesting behaviour of sows and its disturbance by confinement at farrowing. In: Bessei, W. (ed.), *Disturbed Behaviour in Farm Animals*. Verlag Euyn Ulmar, Stuttgart, pp. 101–114.

Benus, R.F. (1988) Aggression and coping. PhD Thesis, University of Groningen, The Netherlands.

Bindra, D. (1959) *Motivation: A Systematic Reinterpretation*. Ronald Press, New York.

Bindra, D. (1976) *A Theory of Intelligent Behavior*. Wiley, New York.

Black, A.J. and Hughes, B.O. (1974) Patterns of comfort behaviour and activity in domestic fowls: a comparison between cages and pens. *British Veterinary Journal* 130, 23–33.

Breland, K. and Breland, M. (1961) The misbehavior of organisms. *American Psychologist* 16, 681–684.

Broom, D.M. and Potter, M.J. (1984) The occurrence of stereotypies in stall-housed dry sows. In: Unshelm, J., van Putten, G. and Zeeb, K. (eds), *Proceedings of the International Congress of Applied Ethology in Farm Animals*. KTBL, Darmstadt, pp. 229–231.

Brouns, F., Edwards, S.A., English, P.R. and Taylor, A.G. (1991) Effects of diet and feeding regime on behaviour of group housed pregnant gilts. In: Appleby, M.C., Horrell, R.I., Petherick, J.C. and Rutter, S.M. (eds), *Applied Animal Behaviour: Past, Present and Future*. UFAW, Potters Bar, UK, pp. 143–144.

Brown, P. and Jenkins, H.M. (1968) Auto-shaping of the pigeon's key-peck. *Journal of the Experimental Analysis of Behavior* 11, 1–8.

Bryant, C.E., Rupniak, N.M.J. and Iversen, S.D. (1988) Effects of different enrichment devices on cage stereotypies and autoaggression in captive Cynomologus monkeys. *Journal of Medical Primatology* 17, 257–269.

Carlstead, K. and Seidensticker, J. (1991) Seasonal variation in stereotypic pacing in an American black bear *Ursus americanus. Behavioural Processes* 25, 155–161.

Carlstead, K., Seidensticker, J. and Baldwin, R. (1991) Environmental enrichment for zoo bears. *Zoo Biology* 10, 3–16.

Cooper, J.J. and Nicol, C.J. (1991) Stereotypic behaviour affects environmental preference in bank voles (*Clethrionomys glareolus*). *Animal Behaviour* 41, 971–977.

Cronin, G.M. (1985) The development and significance of abnormal stereotyped behaviour in tethered sows. PhD Thesis, Agricultural University of Wageningen, The Netherlands.

Cronin, G. and Wiepkema, P.R. (1984) An analysis of stereotyped behaviour in tethered sows. *Annales de Recherches Vétérinaires* 15, 263–270.

Cronin, G.M., van Tartwijk, J.M.F.M., van der Hel, W. and Verstegen, M.W.A. (1986) The influence of degree of adaptation to tether housing by sows in relation to behaviour and energy metabolism. *Animal Production* 42, 257–268.

Dantzer, R. (1986) Behavioral, physiological and functional aspects of stereotyped behavior: a review and a re-interpretation. *Journal of Animal Science* 62, 1776–1786.

Dawkins, M.S. (1983) Battery hens name their price: consumer demand theory and the measurement of ethological 'needs'. *Animal Behaviour* 31, 1195–1205.

Dawkins, M.S. (1988) Behavioural deprivation: a central problem in animal welfare. *Applied Animal Behaviour Science* 20, 209–225.

Dawkins, M.S. (1990) From an animal's point of view: motivation, fitness and animal welfare. *Behavioural and Brain Sciences* 13, 1–61.

de Passillé, A.M.B., Metz, J.H.M., Mekking, P. and Wiepkema, P.R. (1992) Does drinking milk stimulate sucking in young calves? *Applied Animal Behaviour Science* 34, 23–36.

Draper, W.A. and Bernstein, I.S. (1963) Stereotyped behavior and cage size. *Perceptual and Motor Skills* 16, 231–234.

Duncan, I.J.H. and Wood-Gush, D.G.M. (1972a) Thwarting of feeding behaviour in the domestic fowl. *Animal Behaviour* 20, 444–451.

Duncan, I.J.H. and Wood-Gush, D.G.M. (1972b) An analysis of displacement preening in the domestic fowl. *Animal Behaviour* 20, 68–71.

Fentress, J.C. (1972) Development and patterning of movement sequences in inbred mice. In: Kiger, J. (ed.), *The Biology of Behavior*. Oregon State University Press, Corvallis, pp. 83–132.

Fentress, J.C. (1973) Specific and non-specific factors in the causation of behaviour. In: Bateson, P.P.G. and Klopfer, P. (eds), *Perspectives in Ethology*. Plenum Press, New York, pp. 155–224.

Fentress, J.C. (1976) Dynamic boundaries of patterned behaviour: interaction and self-organization. In: Bateson, P.P.G. and Hinde, R.A. (eds), *Growing Points in Ethology*. Cambridge University Press, Cambridge, pp. 135–167.

Fraser, D. (1975) The effect of straw on the behaviour of sows in tether stalls. *Animal Production* 21, 59–68.

Gardner, R.A. and Gardner, B.T. (1988) Feed-forward versus feedback: an ethological alternative to the law of effect. *Behavioural and Brain Sciences* 11, 429–493.

Graves, H.B. (1984) Behavior and ecology of wild and feral swine (*Sus scofa*). *Journal of Animal Science* 58, 482–492.

Hediger, H. (1955) *Studies of the Psychology and Behaviour of Captive Animals in Zoos and Circuses*. Butterworth, London.

Hennessy, J.W. and Levine, S. (1979) Stress, arousal and the pituitary-adrenal system: a psychoendocrine hypothesis. *Progress in Psychology and Physiological Psychology* 8, 134–176.

Hinde, R.A. (1970) *Animal Behaviour*, 2nd edition. McGraw-Hill, New York.

Houston, A.I. and Sumida, B. (1985) A positive feedback model for switching between two activities. *Animal Behaviour* 33, 315–325.

Hughes, B.O. and Duncan, I.J.H. (1988) The notion of ethological 'need', models of motivation, and animal welfare. *Animal Behaviour* 36, 1696–1707.

Hughes, B.O. and Wood-Gush, D.G.M. (1973) An increase in activity of domestic fowls produced by nutritional deficiency. *Animal Behaviour* 21, 10–17.

Hughes, B.O., Duncan, I.J.H. and Brown, M.F. (1989) The performance of nest building by domestic hens: is it more important than the construction of a nest? *Animal Behaviour* 37, 210–214.

Jensen, P. (1988) Diurnal rhythm of bar-biting in relation to other behaviour in pregnant sows. *Applied Animal Behaviour Science* 21, 337–346.

Kandel, E.R. (1979) *Behavioral Biology of Aplysia*. Freeman, San Francisco.

Kastelein, R.A. and Wiepkema, P.R. (1989) A digging trough as occupational therapy for Pacific walruses (*Odobenus rosmarus divergens*) in human care. *Aquatic Mammals* 15, 9–17.

Keiper, R. (1969) Causal factors of stereotypies in caged birds. *Animal Behaviour* 17, 114–119.

Kerr, S.G.C. and Wood-Gush, D.G.M. (1987) The development of behaviour patterns and temperament in dairy heifers. *Behavioural Processes* 15, 1–16.

Kiley-Worthington, M. (1977) *Behavioural Problems of Farm Animals*. Oriel Press, London.

Killeen, P.R., Hanson, S.J. and Osborne, S.R. (1978) Arousal: its genesis and manifestation as response rate. *Psychological Reviews* 85, 571–581.

Kooijman, J., Wierenga, H.K. and Wiepkema, P.R. (1991) Development of abnormal oral behaviour in group-housed veal calves: effects of roughage supply. In: Metz, J.H.M. and Groenstein, C.M. (eds), *New Trends in Veal Calf Production*. Pudoc, Wageningen, pp. 54–58.

Kostal, L., Savory, C.J. and Hughes, B.O. (1992) Diurnal and individual variation in behaviour of restricted-fed broiler breeders. *Applied Animal Behaviour Science* 32, 361–374.

Krzack, W.E., Gonyou, H.W. and Lawrence, L.M. (1991) Wood chewing by stabled horses: diurnal pattern and effects of exercise. *Journal of Animal Science* 69, 1053–1058.

Lam, K., Rupniak, N.M.J. and Iversen, S.D. (1991) Use of a grooming and foraging substrate to reduce cage stereotypies in macaques. *Journal of Medical Primatology* 20, 104–109.

Lawrence, A.B., and Illius, A.W. (1991) Methodology for measuring hunger and food needs using operant conditioning in the pig. *Applied Animal Behaviour Science* 24, 273–285.

Lawrence, A.B. and Terlouw, E.M.C. (1993) Behavioral factors contributing to the development and continued performance of stereotypic behavior. *Journal of Animal Science* (in press).

Lawrence, A.B., Appleby, M.C. and Macleod, H.A. (1988) Measuring hunger in the pig using operant conditioning: the effect of food restriction. *Animal Production* 47, 131–137.

Lawrence, A.B., Appleby, M.C., Illius, A.W. and Macleod, H.A. (1989) Measuring hunger in the pig using operant conditioning: the effect of dietary bulk. *Animal Production* 48, 213–220.

Lawrence, A.B., Terlouw, E.M.C. and Illius, A.W. (1991) Individual differences in behavioural responses of pigs exposed to non-social and social challenges. *Applied Animal Behaviour Science* 30, 73–86.

Lawrence, A.B., Terlouw, E.M.C. and Kyriazkis, I. (1993) The behavioural effects of undernutrition in confined farm animals. *Proceedings of the Nutrition Society* (in press).

Le Magnen, J. (1985) *Hunger*. Cambridge University Press, Cambridge.

Lindsley, D.B. (1960) Attention, consciousness, sleep and wakefulness. In: Field, J., Magoun, H.W. and Hall, V.F. (eds), *Handbook of Physiology. Section 1. Neurophysiology*. American Physiological Society, Washington DC, pp. 1553–1593.

Line, S.W., Markowitz, H., Morgan, K.N. and Strong, S. (1989a) Evaluation of attempts to enrich the environment of singly-caged non-human primates. In: Driscoll, J. (ed.), *Animal Care and Use in Behavioral Research: Regulations, Issues, and Applications*. Animal Welfare Information Center, National Agricultural Library, Beltsville, Maryland, pp. 103–117.

Line, S.W., Morgan, K.N., Markowitz, H. and Strong, S. (1989b) Influence of cage size on heart rate and behavior in rhesus monkeys. *American Journal of Veterinary Research* 50, 1523–1526.

Line, S.W., Morgan, K.N., Markowitz, H. and Strong, S. (1990) Increased cage size does not alter heart rate or behavior in female rhesus monkeys. *American Journal of Primatology* 20, 107–113.

Lorenz, K. (1981) *The Foundations of Ethology*. Springer, New York.

Ludlow, A.R. (1980) The evolution and simulation of a decision-maker. In: Toates, F.M. and Halliday, T.R. (eds), *The Analysis of Motivational Processes*. Academic Press, London, pp. 273–296.

Marsden, D. and Wood-Gush, D.G.M. (1986) A note on the behaviour of individually penned sheep regarding their use for research purposes. *Animal Production* 42, 157–159.

Mason, G. (1991) Stereotypies: a critical review. *Animal Behaviour* 41, 1015–1037.

McFarland, D.J. (1965) Hunger, thirst and displacement pecking in the Barbary Dove. *Animal Behaviour* 13, 293–300.

McFarland, D.J. (1966) On the causal and functional significance of displacement activities. *Zeitschrift fur Tierpsychologie* 23, 217–235.

McFarland, D.J. (1969) Mechanisms of behavioural disinhibition. *Animal Behaviour* 17, 238–242.

McFarland, D.J. (1971) *Feedback Mechanisms in Animal Behaviour*. Academic Press, London.

McFarland, D.J. (1976) Form and function in the temporal organization of behaviour. In: Bateson, P.P.G. and Hinde, R.A. (eds), *Growing Points in Ethology*. Cambridge University Press, Cambridge, pp. 55–93.

McFarland, D.J. (1985) *Animal Behaviour*. Pitman, London.

McFarland, D. (1989) *Problems of Animal Behaviour*. Longman, Harlow, UK.

McFarland, D.J. and Houston, A.I. (1981) *Quantitative Ethology: The State Space Approach.* Pitman Books, London.

Meyer-Holzapfel, M. (1968) Abnormal behaviour in zoo animals. In: Fox, M.W. (ed.), *Abnormal Behaviour in Animals.* Saunders, London, pp. 476–503.

Morgan, M.J. (1974) Resistance to satiation. *Animal Behaviour* 22, 449–466.

Morris, D. (1964) The response of animals to a restricted environment. *Symposium of the Zoological Society of London* 13, 99–118.

Nicol, C.J. (1987) Behavioural responses of laying hens following a period of spatial restriction. *Animal Behaviour* 35, 1709–1719.

Nicol, C.J. and Guilford, T. (1991) Exploratory activity as a measure of motivation in deprived hens. *Animal Behaviour* 41, 333–341.

Ödberg, F. (1978) Abnormal behaviours (stereotypies). In: *Proceedings of the 1st World Congress on Ethology Applied to Zootechnics.* Industrias Graficas Espana, Madrid, pp. 475–480.

Ödberg, F. (1986) The jumping stereotypy in the bank vole (*Clethrionomys glareolus*). *Biology of Behaviour* 11, 130–143.

Ödberg, F.O. (1987a) The influence of cage size and environmental enrichment on the development of stereotypies in bank voles (*Clethrionomys glareolus*). *Behavioural Processes* 14, 155–173.

Ödberg, F. (1987b) Behavioural responses to stress in farm animals. In: Wiepkema, P.R. and van Adrichem, P.W.M. (eds), *The Biology of Stress in Farm Animals.* Martinus Nijhoff, Dordrecht, pp. 135–149.

Palya, W.L. and Zacny, J.P. (1980) Stereotyped adjunctive pecking by caged pigeons. *Animal Learning and Behavior* 8, 293–303.

Pring-Hill, A.F. (1979) Tolerable feedback: a mechanism for behavioural change. *Animal Behaviour* 27, 226–236.

Robert, S., Matte, J.J., Girard, C., Farmer, C. and Martineau, G.P. (1992) Influence de régimes a haute teneur en fibres sur le développement des comportements stéréotypés chez la truie gravide. *Journées de la Recherche Porcine en France* 24, 201–206.

Roper, T.J. (1980) 'Induced' behaviour as evidence of non-specific motivational effects. In: Toates, F.M. and Halliday, T.R. (eds), *Analysis of Motivational Processes.* Academic Press, London, pp. 221–242.

Roper, T.J. (1984) Response of thirsty rats to absence of water: frustration, disinhibition or competition? *Animal Behaviour* 32, 1225–1235.

Rowell, C.H.F. (1961) Displacement grooming in the chaffinch. *Animal Behaviour* 9, 38–63.

Rushen, J. (1984) Stereotyped behaviour, adjunctive drinking and the feeding periods of tethered sows. *Animal Behaviour* 32, 1059–1067.

Rushen, J. (1985) Stereotypies, aggression and the feeding schedules of tethered sows. *Applied Animal Behaviour Science* 14, 137–147.

Rushen, J. (1993) The 'coping' hypothesis of stereotypic behaviour. *Animal Behaviour* (in press).

Sambraus, H.H. (1985) Mouth-based anomalous syndromes. In: Fraser, A.F. (ed.), *Ethology of Farm Animals.* Elsevier, Amsterdam, pp. 381–411.

Savory, C.J., Seawright, E. and Watson, A. (1992) Stereotyped behaviour in broiler breeders in relation to husbandry and opioid receptor blockade. *Applied Animal Behaviour Science* 32, 349–360.

Schouten, W. and Rushen, J. (1992) Effects of naloxone on stereotypic and normal behaviour of tethered and loose-housed sows. *Applied Animal Behaviour Science* 33, 17–26.

Schouten, W. and Wiepkema, P.R. (1991) Coping styles of tethered sows. *Behavioural Processes* 25, 125–132.

Stolba, A., Baker, N. and Wood-Gush, D.G.M. (1983) The characterisation of stereotyped behaviour in stalled sows by informational redundancy. *Behaviour* 87, 157–181.

Terlouw, E.M.C., Lawrence, A.B. and Illius, A.W. (1991a) Influences of feeding level and physical restriction on development of stereotypies in sows. *Animal Behaviour* 42, 981–992.

Terlouw, E.M.C., Lawrence, A.B. and Illius, A.W. (1991b) Relationship between agonistic behavior and propensity to develop excessive drinking and chain manipulation in pigs. *Physiology and Behavior* 50, 493–498.

Terlouw, E.M.C., Wiersma, A., Lawrence, A.B. and Macleod, H.A. (1993a) Ingestion of food specifically facilitates the performance of stereotypies in sows. *Animal Behaviour* (in press)

Terlouw, E.M.C., Lawrence, A.B. and Illius, A.W. (1993b) Relationship between amphetamine and environmentally-induced stereotypies in sows. *Pharmacology, Biochemistry and Behaviour* 43, 347–355.

Timberlake, W. (1983) Rats responses to a moving object related to food or water. *Animal Learning and Behaviour* 11, 309–320.

Toates, F. (1986) *Motivational Systems*. Cambridge University Press, Cambridge.

Toates, F. and Jensen, P. (1991) Ethological and psychological models of motivation – toward a synthesis. In: Meyer, J.A. and Wilson, S. (eds), *From Animals to Animats*. MIT Press, Cambridge, Massachusetts, pp. 194–205.

Vanderwolf, C.H. and Robinson, T.E. (1981) Reticulo-cortical activity and behaviour: a critique of the arousal theory and a new synthesis. *The Behavioural and Brain Sciences* 4, 459–514.

Van Liere, D.W. (1991) Function and organization of dustbathing in laying hens. PhD Thesis, Agricultural University, Wageningen.

Vestergaard, K. (1982) Dustbathing in the domestic fowl: diurnal rhythm and dust-deprivation. *Applied Animal Ethology* 8, 487–495.

Von Borell, E. and Hurnik, J.F. (1991) Stereotypic behavior, adrenocortical function, and open-field behavior of individually confined gestating sows. *Physiology and Behavior* 49, 709–713.

Wemelsfelder, F. (1990) Boredom and laboratory animal welfare. In: Rollin, B.E. and Kesel, M.L. (eds), *The Experimental Animal in Biomedical Research, Vol 1*. CRC Press, Boca Raton, pp. 243–272.

Wiepkema, P.R. (1971) Positive feedbacks at work during feeding. *Behaviour* 39, 266–273.

Wiepkema, P.R. (1985) Abnormal behaviours in farm animals: ethological implications. *Netherlands Journal of Zoology* 35, 279–299.

Wiepkema, P.R. (1987) Behavioural aspects of stress. In: Wiepkema, P.R. and van Adrichem, P.W.M. (eds), *The Biology of Stress in Farm Animals*. Martinus Nijhoff, Dordrecht, pp. 113–133.

Wiepkema, P.R. (1990) Stress: ethological implications. In: Puglisi-Allegra, S. and Oliverio, A. (eds), *Psychobiology of Stress*. Kluwer, Dordrecht, pp. 1–13.

Willard, J.F., Willard, J.C., Wolfram, S.A. and Baker, J.P. (1977) Effect of diet on cecal pH and feeding behavior of horses. *Journal of Animal Science* 45, 87–93.

Wood-Gush, D.G.M. (1972) Strain differences in response to sub-optimal stimuli in the fowl. *Animal Behaviour* 20, 72–76.

Wood-Gush, D.G.M. and Vestergaard, K. (1989) Exploratory behavior and the welfare of intensively kept animals. *Journal of Agricultural Ethics* 2, 161–169.

The Concept of Animal Boredom and its Relationship to Stereotyped Behaviour

FRANÇOISE WEMELSFELDER
Institute of Theoretical Biology, Leiden, The Netherlands.

Editors' Introductory Notes:
No subject produces quite as much division between behavioural scientists as the issue of animal subjectivity. In contrast to the other chapters in this book, Wemelsfelder approaches stereotypies from the perspective of animal subjectivity arguing that models of animal behaviour in general, by failing to deal adequately with subjectivity, cannot be used to infer relationships between development of stereotypies and animal suffering.

The central point of Wemelsfelder's argument is that animals (even quite primitive ones), if given the opportunity, will voluntarily (spontaneously) interact with the environment. For Wemelsfelder this behaviour implies that animals are not simply 'passive' reactors to stimuli but that higher level integrative processes transform animals into organisms that can anticipate events with flexible behavioural responses (i.e. animals seek 'active control' over the environment). Wemelsfelder argues that anticipatory responses cannot be wholly explained in mechanistic terms, and signify the subjective nature of animal behaviour (for example, that animals are aware of 'psychological time'). The operational definition of subjectivity equating to voluntary or anticipatory behaviour allows subjectivity to be investigated in a number of ways, including orienting behaviour, exploration and play and neurophysiological correlates.

Applying this to stereotypies, Wemelsfelder suggests that the nature of intensive housing, by preventing much species-typical behaviour, gradually impairs the animal's capacity to show anticipatory responses giving rise to inflexible (stereotyped) behaviour that is increasingly governed by the immediately available environmental stimuli. The different stages of this process take the animal through 'frustration' (when anticipatory processes are still intact), 'boredom' (growing impairment of these processes) and finally 'depression/anxiety' (when all attempts at active control cease).

Wemelsfelder finally takes issue with the assumption that the development of stereotypies renders the animal an emotionless automaton. She argues alternatively that the decline in active control (expressed as increasingly stereotyped behaviour) indicates an impairment of the subjective integrity of the animal and as such provides direct evidence of animal suffering.

Introduction

Animals which live in natural or semi-natural conditions continuously engage in interaction with their environment. They may show various forms of appetitive behaviour or they may engage in explorative and/or social behaviour. Domesticated animals, when provided with an adequate environment, do not differ significantly from their wild ancestors in this respect (Boice, 1981; Stolba and Wood-Gush, 1989; Vastrade, 1986). However, impoverished housing conditions severely restrict the opportunity to interact with the environment. The animal has little space to move and environmental stimuli which facilitate the expression of species-specific behaviour are mostly lacking. Under such conditions, normal interaction can be gradually replaced with abnormal behaviour patterns such as stereotyped and apathetic behaviour (Dantzer, 1986).

To interpret such abnormal patterns, explanatory concepts such as 'boredom' are frequently used in the experimental literature (see Wemelsfelder, 1990, for a review). However, a model which explains the relationship between boredom and the development of abnormal (notably stereotyped) behaviour has yet to be provided. The term boredom intuitively suggests that animals may suffer from a chronic lack of opportunities to interact (i.e. from having 'nothing to do'). To relate this term to stereotyped behaviour implies that such behaviour is not regarded by the animal as 'doing something', as 'active' in a way comparable to normal species–specific behaviour. In this chapter, I will discuss evidence indicating that stereotyped behaviour reflects the impairment of the active character of behaviour, of the animal's ability to determine the effect of environmental stimulation upon its own behaviour. Behaviour then comes to be largely determined by immediately available environmental stimuli and acquires a passive character. I will argue that the transition from an active to a passive mode of behaviour is initially experienced by the animal as boredom, and subsequently as depression and/or anxiety in the final stages.

Current models of animal behaviour and animal welfare are not likely to interpret such a transition in the organization of behaviour as abnormal. These models generally regard behaviour as a 'black box' in which environmental input is transformed into motoric output by means of control-mechanisms which 'underlie' overt behaviour (see Chapter 3). Such mechanisms presumably have been generated in the evolutionary and individual history of an animal (McFarland, 1989). In such models, the notion of 'control' designates the law-like (i.e. mechanistic) relationship between specific stimuli and behaviour. The causal effectiveness of specific stimuli on behaviour is regarded as fundamentally 'automatic'. Thus, behaviour is regarded as inherently passive. In such a model it is difficult to describe the abnormal character of stereotyped behaviour, because such behaviour presumably is 'controlled by' specific stimuli as well. Abnormal forms of interaction will necessarily be conceived as continuous with normal forms of interaction. As a consequence, criteria which are used to interpret stereotyped behaviour in terms of suffering may seem rather arbitrary (cf. Mason, 1991).

To do justice to the abnormal character of stereotyped behaviour, a model of animal behaviour which acknowledges the active nature of normal species-specific behaviour is needed.* Evidence exists which indicates that through various forms of behavioural orientation, animals actively determine the causal effectiveness of specific stimuli on their own behaviour. This implies that the effect of specific stimuli on behaviour is not automatic. The causal effectiveness of integrative processes prevails over that of the individual stimuli, endowing behaviour with an anticipatory, flexible character (as, for example, in exploration and play). Such control cannot be understood in mechanistic terms. I propose that the anticipatory character of behaviour as manifest in behavioural orientation signifies the subjective character of behaviour. This implies that animal subjectivity is conceived as an *overt* phenomenon. Such a view does not correspond with the way most behavioural scientists interpret subjectivity, namely as an internal, principally unobservable aspect of behaviour. And indeed, the private, experiential aspects of subjective processes are of course unobservable per definition. But the existence of such processes in other organisms need not be unobservable (cf. Griffin, 1976; Dawkins, 1990). I suggest that normally active behaviour directly expresses a state of general subjective integrity or well-being. Stereotyped behaviour may then be conceived as 'abnormal' in that its passive character reflects the impairment of the subjective integrity of the animal and as such provides direct evidence of suffering.

In such a framework, well-being is conceived as a dynamic, not static, notion. This has two important consequences. First, a dynamic notion of animal welfare enables us to investigate well-being through the observation of overt behaviour patterns (i.e. it is 'operational') and therefore may enhance

---

*The distinction between active and passive modes of behavioural causation does not coincide with the distinction between active and passive styles of coping (e.g. Benus, 1988; Schouten and Wiepkema, 1991). The latter distinction refers to individual differences in physiological stress-response and as such falls within a mechanistic interpretation of behaviour; the term 'active', as in active coping, therefore has a passive connotation with regard to behavioural causation. This applies to the notion of 'arousal' as well.

the scientific validity of welfare research (cf. Dawkins, 1990). Second, such a notion enables us to hypothesize that animals, although managing to maintain homeostasis under impoverished housing conditions, may nevertheless suffer from such conditions.

In the first part of this Chapter, I will argue that the concept of animal subjectivity may be defined as the capacity to interact with the environment and that various forms of behavioural orientation may be regarded as a direct parameter of such a capacity. In the second part, I will outline the hypothesis that stereotyped behaviour signifies the impairment of the capacity to interact and as such provides direct evidence of serious suffering. I will first indicate in which way this hypothesis relates to current models of stereotyped behaviour. Subsequently, I will discuss evidence which supports the proposed hypothesis. Finally, I will propose a model of animal boredom and depression which may facilitate a more systematic investigation of these forms of animal suffering.

Defining Animal Subjectivity

ANIMAL SUBJECTIVITY AS THE CAPACITY TO INTERACT WITH THE ENVIRONMENT

In order to argue that an animal's capacity to interact with the environment provides direct evidence of the subjective nature of its behaviour, I will first discuss the problems which surround the view that subjective processes are unobservable in principle. This idea can be regarded as a legacy of 17th century Cartesian dualism, which defines subjective awareness as 'introspection', or the capacity to think. Descartes conceived of the subjective and objective aspects of behaviour as mutually exclusive principles of organization. He postulated that the organization of behaviour is, like that of other physical processes, basically mechanistic (i.e. subject to law-like causal relationships). In contrast, the organization of subjective processes ('thinking') was conceived as non-physical and non-mechanistic. In such a framework, subjectivity is necessarily conceived as 'internal' and unobservable in that it is not causally related to the postulated mechanistic organization of overt behaviour.

Behaviourism has rejected the use of subjective concepts in models of animal behaviour by arguing that the postulation of unobservable behavioural variables does not contribute to the scientific measurement of animal behaviour. It asserts that behavioural causation (i.e. 'reinforcement') takes place in the external environment and not in the animal; the postulation of 'internal' factors is considered to be a self-fulfilling prophecy (Skinner, 1984a,b). However, it must be realized that behaviouristic models do not provide criteria to assess *non*-internalistic, notions of subjectivity. The external, mechanistic nature of behaviour is regarded as given, and as necessarily incompatible with subjective models of behavioural causation. Given such philosophical presumptions, the behaviouristic stance runs the risk of becoming a self-fulfilling prophecy as well.

It has become clear over the past few decades that behaviourism has considerable difficulty in accounting for the active, flexible nature of the behaviour which animals show in complex (semi-)natural settings. To deal with such shortcomings, cognitive models of behaviour now postulate emergent levels of information-processing in the central nervous system; so-called 'internal representations' (e.g. Roitblat, 1982). However, Skinner (1984b) has argued that modern cognitive frameworks represent a form of Cartesian 'mentalism' as well. Present-day behaviourists continue to point out that cognitive models do not provide a viable alternative to behaviouristic models in explaining the active nature of behaviour (Branch, 1982; Catania, 1982; Epstein, 1982, all commenting on Roitblat, 1982). Psychologists as well have argued that in postulating the 'internal' nature of subjective processes, cognitive models do not come any nearer to explaining subjectivity than Cartesian models (Bindra, 1984; Graumann and Sommer, 1984). Bindra (1984), for example, comments that 'each [cognitive] function is left as a self-sufficient agent or transformational homunculus without any attempt to explain it: an "attentional" filter excludes or attenuates messages, a "comparator" compares, a "scanner" scans, and a "memory store" stores'. Thus, it appears that cognitive models do not solve the problem of subjectivity, they restate it.

I suggest that the on-going bickering between cognitivism and behaviourism is a direct reflection of the irreconcilability between 'internal' and 'external' modes of behavioural causation posed by Cartesian dualism. This distinction, however, is not given fact of life. The explanatory problems generated by Cartesian dualism may be avoided by models of subjectivity which do not *a priori* postulate the mechanistic character of animal behaviour and the concomitant unobservability of subjective processes.

Different schools of philosophy have suggested that the Cartesian model does not provide an adequate interpretation of subjective phenomena. They argue that subjective concepts designate a mode of behavioural organization which is directly visible in overt patterns of behaviour, but which cannot be explained in mechanistic terms.

Philosophers of language such as Ryle (1949) assert that Cartesian models do not do justice to our common-sense, every-day experience of subjectivity as an overtly observable aspect of the behaviour of other organisms. Ryle (1949) argues that subjective terms denote states of behavioural competence, that is the ability to deal with the world through behavioural interaction. Behaviour may be performed efficiently, carelessly, skilfully or mindlessly, that is with or without *attention*. A person is designated as intelligent by reference to his or her overt attentive abilities: 'To be intelligent is not merely to satisfy criteria, but to apply them; to regulate one's actions and not to be well-regulated' (p. 29). In other words, subjective concepts denote an active, attentional mode of behaviour. Ryle states that such an interpretation of subjective concepts implies that it is not possible to give an exclusively mechanistic account of overt behaviour. 'Men are not machines, not even ghost-ridden machines. They are men ... a tautology which is sometimes worth remembering' (p. 79). Just so, animals are not machines, even when they are not ghost-ridden machines. Ryle often is regarded as a behaviourist because he argues that subjective concepts describe overt aspects of behaviour; however, his analysis actually points towards a conclusion opposite to that of behaviourism, namely that subjective concepts are necessary to describe the active character of behaviour.

Philosophers in phenomenological tradition also propose a non-Cartesian interpretation of the subjective aspects of behaviour (Buytendijk, 1938; Merleau-Ponti, 1942; von Weizsäcker, 1967). Phenomenological models regard the subjective aspects of behaviour, like other aspects, as a 'phenomenon', that is as overtly observable in principle. Subjective phenomena are supposedly manifest in the attentional abilities of an organism. These abilities endow behaviour with an active, anticipatory character. It is said that to explain these aspects of behaviour, it is necessary to postulate a non-mechanistic causal principle, which is denoted as 'psychological time'. Behaviour at this level of organization expresses an inherent 'future-directedness': past, present and future coexist in every instant. Such future-directedness is regarded as a direct expression of the 'self-initiated' or active character of behaviour (von Weizsäcker, 1967). Buytendijk (1953) has experimentally illustrated the active nature of behavioural organization. In his paper 'Toucher et être touché', he reports that several invertebrate species (e.g. *Aphysia*, *Octopus*) learned to distinguish self-initiated touching from imposed touching. Buytendijk suggests that this ability is a direct expression of the subjective character of animal behaviour.

Mechanistic models of behaviour essentially postulate that behavioural organization takes place in 'physical time'; that is as a sequence of discrete, unrelated 'moments' of causation. Descartes conceived of the natural world as a gigantic clock set in motion by God. It is therefore not surprising that the explanation of 'psychological time' is considered a major problem in the physical sciences (Davies and Gribbin, 1991). Some physicists (e.g. Hawking, 1988) assert that psychological time is nothing but the 'arrow of time' generated by the inherent tendency of the universe towards increased entropy. Such a tendency towards chaos is explained by the second law of thermodynamics, and does not require the postulation of psychological factors. However, other physicists argue that such a notion of the 'arrow of time' does not do justice to the relationship between past, present and future which is experienced in the 'here and now'. 'The idea that time "flows", or that the present moment somehow moves from the past to the future in time, has no place in the physicist's description of the world' (Davies and Gribbin, 1991, p. 128). So far, the irreducibly subjective nature of psychological time has not been repudiated by modern physical theories.

Animals presumably experience psychological time as a sense of 'having a future', that is as a feeling of general psychological vitality and meaningfulness. Such an interpretation of subjectivity does not refer to invisible processes of introspection and/or thinking. It refers to the experience of 'relationship with', or 'existence in', one's environment (cf. Heidegger, 1927). A related interpretation of animal subjectivity has been proposed by Nagel (1974) in his famous paper 'What is it like to be a bat?'. Nagel argues that subjective concepts denote the experiential, private aspects of the relationship which an animal acquires with the environment through its species-specific behaviour.

As complicated as the notion of 'psychological time' may seem, I believe that it is essential to understand the problem of animal boredom. Intuitively, it seems that a feeling of boredom concerns the way we experience the passage of time. Introducing the concept of 'psychological time' is the only way, I propose, in which it can be argued that animals experience the passage of physical time as meaningless. I will come back to this later when I propose a model of animal boredom and depression.

In summary, the concept of animal subjectivity may be defined as the capacity to interact with the environment in that this capacity endows behaviour with an active, future-directed nature which cannot be explained in mechanistic terms.

PARAMETERS FOR THE INVESTIGATION OF ANIMAL SUBJECTIVITY

Introduction

Orienting behaviour, exploration and play may be regarded as a direct expression of the capacity to interact with the environment. These behaviours reflect an animal's ability to direct attention towards the environment and to anticipate novel stimulation. As such, they are indicative of the active nature of behavioural organization. Current models of animal behaviour deal with behavioural anticipation by postulating concepts such as 'feed-forward' (Gardner and Gardner, 1988), 'reafference' (von Holst and Mittelstaedt, 1950; Held, 1961) and 'synchronicity' (Gallistel, 1990). Such concepts are not endowed with subjective connotation; it is assumed that anticipation is a property of in-built stimulus-control mechanisms which serve to update neurally encoded information (McFarland, 1989).

However, such models do not do justice to the innovative nature of behavioural learning. Behavioural learning does not just imply that existing neural information is 'updated'. Moreover, it implies that existing information is actively *used* by the animal to develop new behavioural strategies in dealing with novel aspects of the environment; an ability which may be denoted as behavioural flexibility (Fagen, 1982). Evidence exists that active, self-produced movement is a prerequisite for the adaptive use of information in a visual learning task (Held and Freedman, 1963; Held and Hein, 1963). If young kittens and human adults were exposed to visual stimuli by means of imposed, passive movement only, sensorimotor coordination in a learning task which involved these stimuli became seriously disturbed. The authors conclude that 'although the passive-movement condition provided the eye with the same optical information that the active-movement condition did, the crucial connection between motor output and visual feedback was lacking' (Held and Freedman, 1963). Such data indicate that behavioural learning does not occur automatically, but is an active ability. A model of motivation and learning proposed by Mowrer (1960) explicitly deals with the active aspects of learning and suggests that they provide direct evidence of 'volition'. Orienting behaviour presumably enables an animal to bring 'the remote and immediate consequences of an impending overt action . . . into the psychological present' (Mowrer and Ullman, 1945). In other words, through behavioural orientation, an animal exists in 'psychological time'.

An important consequence of the active character of behavioural learning is that learning should not be dependent upon the actual presence of stimuli. Animals should actively bring about change in the environment, or search for stimuli which might be of interest to them. Such a tendency will increase the general diversity and versatility of behaviour and enhance the animal's competence in dealing with the environment (White, 1959). In this section, I will discuss evidence which indicates that orientative, explorative and playful behaviours serve to solve novel problems and to expand the animal's behavioural repertoire, and may therefore be regarded as a direct expression of animal subjectivity.

Orienting behaviour

Orienting behaviour consists of movements of the head and/or the whole body which direct the sensory organs towards a perceived goal or stimulus. In learning experiments, such movements occur mostly at points in which a decision between various behavioural options must be made. The animal alternately orients itself towards the different stimuli, as if it were trying to decide what to do. Tolman (1948) has defined orienting behaviour as the 'hesitating, looking back and forth sort of behaviour which rats can often be observed to indulge in at a choice point before actually going one way or the other'.

Studies of orienting behaviour are largely restricted to the early part of this century (see Goss and Wischner, 1956, for a review). These studies were mostly concerned with mammals, but orienting behaviour has since been described as well in birds (Daanje, 1951), sticklebacks (Sevenster and Roosmalen, 1985) and even in as 'primitive' a species as the planarian (Best, 1963).

Thinus-Blanc (1988) is one of the few scientists who has recently discussed the importance of self-initiated exploratory movement for behavioural learning. She points out that to acquire information on the spatial structure of the environment, an animal has to 'manipulate space' by actively moving around.

Krechevsky (1932, 1936) and Tolman (1932, 1948) have investigated the functional importance of orienting behaviour for behavioural learning. Krechevsky (1932) demonstrated that discrimination learning in rats consists of systematic attempts to test out a variety of 'hypotheses'. A detailed analysis of behaviour patterns shown prior to reaching performance criteria, revealed that these patterns were not randomly distributed over the various experimental clues, but formed highly ordered sequences of behaviour. These results were later reproduced in rhesus monkeys (Levine, 1959). Krechevsky (1936) argues that behavioural variability should not be regarded as a by-product of random environmental stimulation (as 'errors'), but as a primary behavioural principle which he designates as 'attentiveness'.

Tolman, (1932, 1948) more explicitly investigated the role of orienting behaviour in the problem-solving of white rats. His results indicate that the frequency of orienting behaviour is correlated to the animal's successfulness in 'catching on' to a problem. Tolman (1948) concludes that 'the animal's activity is not just one of responding passively to discrete stimuli, but rather one of the active selecting and comparing of stimuli'. The performance of orienting behaviour is regarded as evidence of 'means-end-readiness' or 'consciousness' (Tolman, 1932). Tolman's work is generally regarded as an important precursor of cognitive models of animal behaviour. It should be realized, however, that unlike modern models of cognition, Tolman primarily relates the subjective nature of behaviour to *overt* behaviour patterns. As a consequence, he has been made out as a (neo)behaviourist. I suggest that this is unwarranted and that both the work of Krechevsky and of Tolman provide valuable examples as to how the subjective aspects of behavioural learning may be investigated experimentally.

Several studies of orienting behaviour suggest that the active nature of this behaviour may be due to its inherent *rhythmicity*. This may have important consequences for the investigation of subjective states (see below). Welker (1964) and Komisaruk (1970) have demonstrated that orientative sniffing in white rats consists of sequences of very fast, highly synchronous rhythmic movements of head, nostrils and vibrissae. A similar rhythmical organization of sniffing behaviour has been observed in the tree shrew *Tupaia belangeri* (von Holst and Kolb, 1976). Welker (1964) suggests that the inherent variability and synchronicity of the various levels of rhythmic movement facilitate an optimally efficient reception of sensory information. Thelen (1979, 1981a,b) has studied rhythmical motor patterns in human infants. She defines such patterns as 'the quite rapid, repetitious movement of the limbs, torso, and head that is quite common in human infants during the first year or so of life' (Thelen, 1981b). Her observations demonstrate that the overall frequency of such patterns declines as the infant gets older, but that the diversity and variety of rhythmical behaviour gradually increases and merges into more purposive, voluntary patterns of behaviour (Thelen, 1979). Rhythmical movement, it is suggested, facilitates the emergence of 'voluntary control' (Thelen, 1981b). Zelazo and Kearsley (1980) describe a similar transition from 'stereotyped' to 'functional' play.

Thelen (1981b) speaks of 'rhythmical stereotypes' because of the seemingly repetitive nature of rhythmic movement. However, in contrast to stereotyped behaviour in captive animals (see p. 80), rhythmic movement in infants facilitates the development of behavioural variability and diversity (Berkson, 1983). It often occurs in close connection to exploration and play (Hutt, 1970; Hughes, 1983). This indicates that rhythmical movement serves to anticipate novel aspects of behaviour. It should therefore not be regarded as repetitive, but as an expression of the subjective character of behavioural learning (cf. Kuijper, 1963; Butterworth and Hopkins, 1988).

Exploration and play

Behavioural orientation towards the environment does not just occur in response to a perceived problem or novel object. Interaction with the environment may become an end in itself, as is evident in the performance of explorative and playful behaviours. The animal's tendency to actively search for changes in the environment has been denoted as 'inquisitive' exploration, in contrast to a more passive, 'inspective' form of exploration in which the animal is confronted with a novel stimulus (Berlyne, 1960; Russell, 1983). In play, the animal's attention is directed towards changes which it produces through its own movement. A clear distinction between inquisitive exploration and play is often difficult to make (Hughes, 1983). The dynamic, open-ended quality of both types of activity suggests that they directly reflect the subjective nature of animal behaviour.

Experimental approaches to exploration have mostly focused on inspective exploration. Ample evidence is available that environmental cues such as 'stimulus-change' (e.g. Kish, 1955) 'stimulus-novelty' (e.g. Welker, 1956) or 'stimulus-complexity' (e.g. Dember et al., 1957) elicit an immediate explorative response. Inspective exploration appears to be regulated by homeostatic mechanisms: the explorative response habituates as the novelty of the situation decreases. Current models of exploration interpret such data in terms of neural information-processing, by postulating that organisms strive to reduce the uncertainty which is induced by novel stimulation (Birke and Archer, 1983; Inglis, 1983).

Homeostatic models of exploration have difficulty in accounting for inquisitive exploration. Animals continue to search for novel stimulation or to manipulate a novel object much longer than would be expected by such models (Harlow, 1950; Harlow et al., 1950; Butler and Harlow, 1954; Butler and Alexander, 1955, for primates; Hutt, 1966, for children; Stahl et al., 1973, for pigeons). Some models nevertheless attempt to explain inquisitive exploration in homeostatic terms. It is postulated that 'higher-order' set-points exist which concern the level of environmental change itself; the animal may become understimulated and will start to actively seek information (e.g. Inglis, 1983). However, such a model still has difficulty in explaining the persistence with which inquisitive exploration occurs, not just in unchanging environments, but in normally variable environments as well (Harlow, 1950; Wood-Gush et al., 1990). Furthermore, it is questionable whether the postulation of set-points at 'higher' levels of organization really explains anything, or whether it merely restates the problem to be explained. Rather than explaining the active aspects of behaviour by postulating a hierarchy of (passive) homeostatic mechanisms, these aspects may be regarded as a direct reflection of the subjective organization of behaviour.

Inquisitive exploration may merge into various forms of play (Wood-Gush and Vestergaard, 1991). An essential characteristic of play is its versatility; movements are often exaggerated, truncated, and/or repeated several times. Functional units from the basic motivational behaviours are combined with each other and with non-functional units in unpredictable order (Einon, 1983; Martin and Caro, 1985). As a consequence, controlled observation of play is difficult. Nevertheless, several studies indicate that play enhances the flexible and innovative character of behaviour (Fagen, 1982, 1984).

Play may alleviate the detrimental effects of social isolation on the learning ability of white rats. White rats which are socially isolated until they are 45 days old are markedly slower in transfer and reversal of a learning task than rats which are reared in social conditions; this is interpreted as indicating that isolated rats are deficient in behavioural flexibility (Morgan, 1973; Morgan et al., 1975). Providing the rats with the opportunity to play for 1 hour each day significantly reduces such effects of isolation (Einon et al., 1978). The authors suggest that the versatility of play may enhance the development of behavioural flexibility.

Furthermore, play may affect brain plasticity, that is the neural organization of behavioural learning (Ferchmin et al., 1980). White rats which are provided with various forms of sensory stimulation, but are prevented from interacting with the perceived stimuli, do not show an increase in brain weight similar to animals which can freely interact with their environment. Furthermore, these rats make significantly less entries into the arms of a cross-maze than freely interacting animals (Ferchmin et al., 1975). The same has been found in rats which are trained to perform various complex motor acts in an otherwise impoverished environment (Ferchmin and Eterovic, 1977). The authors suggest that a self-initiated, active form of interaction with the environment, such as play, is a predominant factor in brain plasticity, because it facilitates sensorimotor integration. Explicit evidence that the importance of play for learning supersedes that of both perceptive and motoric training was found in human children by Sylva et al. (1974) and Smith and Dutton (1979).

The relationship between play and behavioural innovation has repeatedly been observed in primates (Kummer and Goodall, 1985). Chimpanzees which have time to play with sticks or branches are able to use these as tools in entirely novel ways (e.g. in gaining access to food), whereas animals which are given the sticks but not the opportunity to play, are not (Jackson, 1942; Birch, 1945). In reviewing the experimental literature on play in children, Hutt (1970) and Hughes (1983) both conclude that play facilitates behavioural innovation and flexibility.

Neurophysiological correlates

Research on the neurophysiological correlates of orienting behaviour indicates that the anticipatory organization of behaviour is reflected in rhythmical patterns of electrical activity in the brain (EEG) (Rohrbaugh, 1984). Various forms of hippocampal EEG (e.g. theta) and scalp-recorded EEG are related to the ability of experimental subjects to direct and/or sustain attention to an impending action or a perceived behavioural task (Pribram and McGuinness, 1975; Thompson et al., 1979; Libet, 1985). Such EEG patterns are generally interpreted as signifying a sensorimotor state of 'alertness' (Kahneman, 1973). Such a state presumably plays an important role in the long-term regulation of behaviour; that is, it facilitates behavioural 'sensitization' (Groves and Thompson, 1970) or 'mismatch-detection' (O'Keefe and Nadel, 1978).

Concepts which are used to explain the relationship between such neural states and behaviour are, for example, 'conscious attention' (Posner, 1978), 'voluntary attention' (Pribram and McGuinness, 1975), 'volition' (Libet, 1985), 'effort' and 'concentration' (Kahneman 1973). However, the use of such subjective concepts to explain neurophysiological phenomena is controversial. Many scientists feel that such concepts do not contribute to a scientific explanation of behaviour. Neurophysiological states should explain the 'subjective' aspects of behaviour, and not the other way round.

It is true of course that it would be a mistake to assume that the use of subjective concepts endows neurophysiological states with a subjective connotation. The anticipatory character of orienting behaviour cannot be explained in terms of (i.e. 'reduced to') 'underlying' physical mechanisms (see Introduction). 'Curiosity', for example, is directly visible in the way a bird tilts its head, but not in the EEG patterns which accompany such movements. It is not the EEG pattern, but the bird which is curious. However, this does not discard the possibility that subjective concepts are essential in explaining the dynamics of attention at a neurophysiological level of behaviour. Such concepts designate the highest level of behavioural control and as such necessarily have explanatory validity with regard to physical mechanisms at lower levels of control. It is therefore not unwarranted to postulate that the subjective nature of behaviour is reflected in neurophysiological parameters. I propose that in neurophysiological terms, subjectivity may be defined as the ability to maintain a central state of 'alertness'.

In summary, the subjective aspects of animal behaviour are manifest in the various ways in which an animal orients itself towards the environment. To investigate these aspects, a descriptive framework for orienting behaviour must be developed. A detailed analysis of the timing and direction (i.e. rhythmicity) of behaviour may provide operational criteria for the interpretation of subjective states. An animal may, for example, approach a novel object quickly, hesitantly, directly, or waveringly. Such qualifications of attentional mode may in turn be used to define the subjective state of the animal: the animal's approach may be described as inquisitive, aggressive, fearful, etc. Such states may be accompanied by various EEG parameters.

The ability to anticipate novel and/or unexpected events in the environment has been described for practically all animal species (cf. Griffin, 1976, 1984). All animals may therefore be attributed with subjective states. This does not imply, however, that they are all equally capable of acquiring abstract knowledge. The various species would presumably differ in the *scope* of their subjective perspective, that is in the length of time at which they would be able to anticipate future events. The greater the scope, the less a prevailing stimulus situation would influence current behaviour, and the more 'abstract' or 'context-free' learning would be. Animals with a greater anticipatory scope would learn spatial 'maps' rather than 'routes' (Thinus-Blanc, 1988) and would form 'plans' rather than 'habits' (Dickinson, 1985). In the model of animal subjectivity proposed in this paper, such a distinction does not refer to subjective versus mechanical modes of existence. It refers to different modes of behavioural organization in which a subject can interact with its environment.

The model of animal subjectivity outlined in the first half of this chapter is essential to understand the suffering of animals housed in impoverished conditions. First, it explains why interaction with the environment is a prerequisite to the maintenance of subjective well-being (cf. Hughes and Duncan, 1988). Second, by defining animal subjectivity as an overt phenomenon, it provides criteria to investigate whether or not stereotyped behaviour is an abnormal form of behaviour and as such may be regarded as a direct expression of suffering.

Interpreting Stereotyped Behaviour

EXPLANATORY LIMITATIONS OF CURRENT MODELS OF STEREOTYPED BEHAVIOUR

A model of stereotyped behaviour which acknowledges the subjective nature of animal behaviour complements current models of stereotyped behaviour in essential respects. As outlined in the Introduction, current models of stereotyped behaviour generally interpret behavioural causation in mechanistic, that is passive, terms. Such models are well equipped to describe and interpret the effect of impoverished housing conditions on specific stimulus-control mechanisms (e.g. feeding or nest-building). However, they will leave the effect of impoverished housing conditions on the active organization of behaviour undescribed. It is assumed that behavioural fixation is either functionally equivalent to, or continuous with, normal forms of behavioural control. This makes it difficult to assess the effect of behavioural fixation on well-being in its own right. For example, concepts such as 'self-organization' (Fentress, 1976), 'adaptation' (McBride, 1980), 'de-arousal' (Wood-Gush et al., 1983) and 'coping' (Broom, 1988) are frequently used in the interpretation of stereotyped behaviour and imply the continuity of stereotyped behaviour with normal forms of control. They may therefore be (mis)interpreted as suggesting the absence of suffering. If it is not realized that mechanistic models of animal behaviour are not equipped to interpret changes in the active control of behaviour, considerable confusion may arise with respect to chronic suffering in captive animals. Several examples may illustrate this.

First, models which take 'biological fitness' as their main criterion for animal well-being generally regard behaviour as subservient to the maintenance of physiological homeostasis. Physiological shifts which occur in response to environmental stressors tend to be regarded as a restoration of control, regardless of the abnormal character of accompanying behaviour. For example, Barnett and Hemsworth (1991) argue that because stereotyped behaviour is associated with seemingly adaptive physiological shifts (but see Chapters 6 and 7 for a critical discussion of such an assumption), 'the performance of stereotypies may be regarded as a mechanism that helps animals 'to successfully adapt to conflict or tethering'. In this context, the term 'successfully' refers to those aspects of well-being which are related to the animal's physical survival. However, to say that stereotypies 'help' the animal to cope evokes the unfounded suggestion that stereotypies are continuous with normal forms of behavioural control and generally do not affect welfare. Such an assumption is unwarranted unless the effect of behavioural fixation on well-being is assessed in its own right.

Second, models which describe behavioural organization as an information-processing system currently play an important role in the field of animal welfare. Such models tend not to directly associate long-term shifts in information-processing capacity with suffering. For example, Wiepkema (1987) interprets the development of stereotyped behaviour as a form of 'habit-formation' which is induced by the high predictability of impoverished environments. The term habit-formation implies that stereotypies are conceived as functionally equivalent to normal forms of behavioural control. The predictability of impoverished environments is *a priori* regarded as similar to that of normally variable environments: 'when the same situation presents itself over and over again flexible programs may change into routines, or actions into habits (cf. Dickinson, 1985)' (Wiepkema, 1987, p. 129/130). Wiepkema (1987) concludes that 'routines and habits, because of their high predictable/controllable outcomes, are no longer associated with emotions; organisms may perform such type of behaviour programs practically emotionless' (p. 130). The implications of such a conclusion with regard to suffering are not clear. However, the postulated emotionlessness of stereotyped behaviour cannot be interpreted as implying the general absence of suffering as long as it is unclear to what extent impoverished environments affect active control.

Other models do not assume that stereotyped behaviour is functionally equivalent to normal forms of control (e.g. Dantzer, 1986; Hughes and Duncan, 1988). Hughes and Duncan (1988) suggest that stereotyped behaviour 'has become divorced from its functional consequences' and that the animal 'may get into a closed loop from which it cannot escape'. The authors leave it as an open question whether or not the postulated dissociation between behaviour and its functional consequences involves suffering. Dantzer (1986) proposes 'that the occurrence of stereotypes reflects a cut-off of higher nervous functions ...' However, he suggests that because the organization of stereotypies is 'hard-wired', animals do not suffer while engaged in stereotyped behaviour. Such a view is based upon the *a priori* assumption that given the basically hard-wired, mechanistic nature of behaviour, the impairment of control does not affect the animal subjectively. However, a model which acknowledges the non-mechanistic aspects of behaviour may lead to a different conclusion. The disruption of behavioural control may then be interpreted as direct evidence of chronic suffering.

In the remainder of this chapter, I will propose such a model. Evidence will be discussed which indicates that the development of stereotyped behaviour signifies the gradual impairment of an animal's capacity to interact with the environment (i.e. of active control). I will argue that the chronic suffering which ensues from such behavioural impairment may be interpreted in terms of boredom and depression and/or anxiety.

STEREOTYPED BEHAVIOUR AS AN INDICATION THAT THE CAPACITY TO INTERACT WITH THE ENVIRONMENT IS IMPAIRED

The development of stereotyped behaviour in impoverished housing conditions may be described as a gradual process of behavioural fixation (Stolba et al., 1983; Cronin and Wiepkema, 1984; Dantzer, 1986; Wechsler, 1991). Patterns of short duration and high fixation increasingly dominate behaviour. As a consequence, behaviour becomes increasingly repetitive and is performed at greater speed than normal (Dantzer, 1986). In the long term, self-directed stereotypies tend to replace stereotypies which are directed towards the environment (Chamove and Anderson, 1981; Stolba et al., 1983; Dantzer, 1986). Furthermore, animals which develop stereotyped behaviour show an overall reduction in behavioural diversity (Dantzer, 1986) and may frequently assume immobile bodily postures (Buchenauer, 1981; Stolba et al., 1983; Broom, 1986).

Behavioural fixation has been interpreted as a form of behavioural self-organization (cf. Fentress, 1976). The self-organization of behaviour is a functional response in normally variable environments, which serves to minimize the need to pay attention to on-going activities such as feeding and grooming, thereby freeing the animal to direct its attention to events which require active processing (Fentress, 1976). Self-organized patterns acquire a relatively autonomous character, and may not be easily disrupted by external input. Such characteristics are also typical of the stereotyped patterns shown by animals in impoverished environments. However, normal self-organization is a function of attention and facilitates the overall flexibility of behaviour. Individual elements of behaviour may acquire an autonomous character, but, on the whole, behaviour retains its flexible and variable nature; the animal will continue to respond adequately to novel and/or unexpected situations. The stereotyped patterns which develop in impoverished environments, on the other hand, do not appear to facilitate behavioural flexibility. On the contrary, they appear to reflect the increasingly rigid and *in*flexible character of behaviour. This suggests that behavioural fixation signifies the impairment, rather than the facilitation, of attention. The following evidence supports such an hypothesis.

In some cases, stereotyped patterns largely disappear when the variability and complexity of the animal's environment are increased by means of species-specific and/or novel stimuli (Keiper, 1970; Stevenson, 1983; Ödberg, 1987; Kastelein and Wiepkema, 1989). In other cases, however, stereotyped and self-directed patterns persist in their original form, or even increase, despite the provision of environmental enrichment (Berkson et al., 1963; Hutt and Hutt, 1965; Keiper, 1970; Stevenson, 1983). Such patterns may persist even when the animal is permanently transferred to enriched conditions (Morris, 1964; Meyer-Holzapfel, 1968). Such a failure to integrate previously established patterns into a novel situation indicates that behavioural fixation is not a function of attention, but reflects a growing impairment of an animal's ability to respond flexibly to changing environmental conditions.

A direct correlation between behavioural fixation and a change in responsiveness towards novel stimuli has been demonstrated by Wood-Gush et al. (1983). These authors found that as the level of behavioural fixation in tethered sows increases, the responsiveness of these animals to a dangling sack decreases. Similarly, level of stereotypy in autistic children is inversely related to responsivity to a candy-disperser (as measured by response-latency) (Lovaas et al., 1971). On the other hand, stereotyping animals may also show an overly aggressive and/or fearful reaction towards novel stimuli (Broom, 1986). Animals with established stereotypies do not necessarily become entirely inattentive to novel stimuli; nevertheless, their tendency to interact with environmental stimuli decreases considerably (Wechsler, 1991). Both the observed decrease and increase in responsiveness indicate that stereotyping animals lose their ability to approach a novel stimulus *in an active, functionally meaningful way*. However, the absence of a response to a certain stimulus may simply be due to the animal's lack of interest in that stimulus and not to the impairment of its anticipatory capacities. To distinguish between these possibilities, it is necessary to take the effect of behavioural fixation on other measurements of behavioural anticipation into consideration as well.

That behavioural fixation generally affects an animal's anticipatory capacities is illustrated by a study which has investigated the effect of stereotyped behaviour on environmental preference in bank voles (Cooper and Nicol, 1991). Whereas non-stereotyping voles showed a clear preference for an enriched environment, stereotyping voles divided their time more equally between impoverished and enriched environments. The authors suggest that the stereotyping voles show less preference for the enriched environment because the performance of stereotypies reduces the aversiveness of the barren environment. However, I disagree with this interpretation in that the term 'preference' may be considered inappropriate to explain the behaviour of animals which show high levels of behavioural fixation. I suggest that the behaviour of stereotyping voles is divided more randomly between the two environments because the ability of these animals to anticipate the consequences of their own behaviour has become impaired; as a consequence they have become incapable of exercising consistent choice behaviour.

Second, evidence exists which indicates that the development of stereotypies takes place at the expense of exploration and play, and vice versa. If, for example, captive walruses are given the opportunity to search for their food, stereotyped swimming is reduced and the performance of playful, explorative and orientative behaviours increases (Kastelein and Wiepkema, 1989). The data of this study show that the reduction of stereotyped patterns and the concomitant increase in more flexible patterns of behaviour is not merely due to the new time-budget as induced by the food-seeking device (Terlouw et al., 1991). This suggests that an inverse relationship exists between behavioural fixation and behavioural flexibility as general principles of behavioural organization. Such a relationship has also been observed in contexts other than that of environmental enrichment. Koegel et al. (1974) found that a suppression of stereotyped behaviour in autistic children induced an increase in spontaneous play more than in other activities. The reverse effect was found as well: suppression of play induced an increase in stereotyped behaviour.

On the basis of the evidence discussed above, I propose that the development of stereotyped behaviour in captive animals signifies *the gradual impairment of the capacity to interact with the environment*. Behaviour acquires an increasingly rigid and mechanical character; it is determined more and more by immediately available environmental stimuli and loses its innovative, anticipatory nature. Behavioural fixation and behavioural flexibility should be regarded as opposed principles of organization. It follows that the repetitiveness of stereotyped behaviour is not equivalent to the rhythmicity of normal orientative behaviour (Kuijper, 1963). Instead, it signifies an increasing *in*ability to integrate sensory feedback into adaptive patterns of behaviour; the animal loses its ability for behavioural control. I propose that it is such a process of behavioural disintegration which constitutes the abnormality of stereotyped behaviour, in both its overt and its experiential aspects.

This hypothesis may be investigated by observing the relationship between changes in an animal's capacity to interact with the environment and the development of stereotyped behaviour. Some domesticated species (e.g. guinea-pigs and rats) do not develop stereotypies under impoverished conditions. Within one species, individual differences may occur: some animals may perform very little stereotyped behaviour, instead showing high levels of lying and sitting (e.g. Wechsler, 1991). However, forms of abnormal behaviour other than stereotyped behaviour may also reflect impaired behavioural control. This may be tested as follows.

Orienting behaviour is a direct expression of an animal's capacity to interact with the environment (see p. 72). The long-term effect of impoverished housing conditions on this capacity may therefore be measured by describing changes which occur in the animal's orienting response towards various forms of novel stimulation, which are presented to it at subsequent points in time. The animal may repeatedly be brought into an unfamiliar environment or be provided with novel objects in its own home-cage. It may be expected that the animal's tendency to approach novel stimuli and to engage in some form of interaction with these stimuli will decrease gradually. Concomitant to such changes, exploration and play may gradually decrease as well.

Some evidence for such an effect of impoverished housing conditions upon behaviour already exists. Animals which are raised in impoverished conditions do not manipulate novel objects as enriched conspecifics do; for example, isolated rats do not move or climb objects but investigate them in a less active, more generally explorative fashion (e.g. through sniffing) (Mason and Green, 1962; Einon and Morgan, 1976; Renner and Rosenzweig, 1986). Sachser and Lick (1991) furthermore found that socially deprived male guinea-pigs approach an unfamiliar male more aggressively than enriched ones; it is suggested that the deprived animals have a decreased ability to deal flexibly and strategically with social conflicts. Such studies indicate that impoverished conditions affect the animal's ability for active control; however, to investigate the hypothesis outlined above, a systematic and detailed description of the way impoverished housing conditions affect behavioural orientation is urgently needed.

To interpret observed changes in orientation, a model must be developed which relates these changes to subjective concepts which are used to designate different forms of animal and human suffering. In the final section of this chapter, I will outline such a model.

STEREOTYPED BEHAVIOUR AS AN INDICATION OF CHRONIC SUFFERING

In the previous sections, I have argued that the abnormal, disintegrated character of stereotyped behaviour provides direct evidence of suffering. I propose that the gradual impairment of an animal's ability to interact with the environment may be conceived as *boredom* in the early stages of impairment and as *depression* and/or *anxiety* in later, more severe stages of impairment. These states may arise out of and/or overlap with a state of suffering in which this ability is still largely intact, and which may be designated as *frustration*. The different states may be distinguished experimentally through measuring the animal's orienting response towards a novel stimulus.

Frustration

Current models of stereotyped behaviour have provided ample evidence that the restriction of an animal's specific motivational tendencies is an important factor in the development of stereotypies (for a review, see Chapter 3). The concept of frustration has been used frequently to interpret stereotyped behaviour (Maier, 1949; Duncan, 1970; Duncan and Wood-Gush, 1972; Wood-Gush, 1972; Wood-Gush et al., 1975; Dawkins, 1980; Rushen, 1985) and I agree that frustration is a useful concept to describe those aspects of stereotyped behaviour which are induced by the blockage of specific motivations and which concern specific stimulus-control mechanisms. However, the concept of frustration has an active connotation (cf. Hinde, 1970), which suggests that active control is intact. It is therefore not adequate to interpret the effect of impoverished housing conditions on active control itself.

Frustration may be the predominant form of suffering in the early stages of behavioural fixation, when stereotyped patterns are not yet established. Behavioural control is presumably still largely intact at this stage. This may be tested by investigating whether or not an animal's response towards novel stimuli is normally inquisitive and manipulative. It would be expected that, if adequate substrate to alleviate the animal's frustration is provided, stereotyped patterns will disappear.

Boredom

As it spends more time in impoverished conditions, the animal increasingly directs it behaviour towards inadequate stimuli (e.g. McKinnon et al., 1989). It exaggerates normal patterns of behaviour and stereotyped patterns become more established. Examples of redirected behaviours are tail-biting in pigs (van Putten and Dammers, 1976), feather-pecking in hens (Blokhuis, 1986), self-mutilation in primates (Fox, 1986) and excessive grooming in mice and rats (Militzer and Wecker, 1986). The performance of redirected behaviour may be regarded as a form of behavioural fixation in that such behaviour, like stereotyped behaviour, indicates that the animal is not sufficiently capable of anticipating the environmental consequences of its own behaviour and increasingly responds towards inappropriate, immediately available environmental stimuli. Moreover, redirected patterns may become stereotyped in time (e.g. chain-nibbling in pigs, Stolba et al., 1983).

This stage of behavioural fixation may be described as a state of boredom, in that the animal experiences the growing impairment of its anticipatory capacities as the increasing dissolution of 'psychological time' (see p. 70). Behaviour loses its integrated and flexible character and is less and less experienced as a meaningful process. As the animal's sense of psychological time becomes increasingly empty, the passage of time acquires an increasingly 'physical' character. That is, it is increasingly experienced as repetitive, as lasting longer than it should. I suggest that the enduring lack of opportunities for interaction induces a feeling in the animal that time goes by unused (cf. Barmack, 1938). Thus, the concept of boredom can only be understood in a model which acknowledges the subjective character of overt behaviour. Such a model explains why an animal may suffer from a behavioural state in which behaviour is increasingly determined by immediately available stimuli.

In daily human life, the term boredom is often used in a fairly harmless context. We say we are bored with a situation when we see no more interesting opportunities for interaction and want to leave. However, the term boredom is also used to qualify more harmful pathological conditions, namely the listless state into which individuals may sink when they perceive their situation as perspectiveless. A lack of meaningful direction in one's life generates a sense of empty time, to be filled with compensatory 'excitement'. The potentially pathological nature of such a condition is that the search for compensatory stimulation may become 'addictive'. To fill the time with substitute stimulation does not relieve the experienced lack of inner time. On the contrary, this feeling may become stronger and the individual may get trapped into cycles of frantic replacement behaviour. One may for example become addicted to playing videogames, to gambling, or to various forms of aggressive behaviour, to name a few examples.

Such an interpretation of the term boredom may also be applied to animals whose housing conditions deprive them of opportunities to interact with the environment. Indeed, it is known that redirected behaviours such as tail-biting and feather-pecking are very hard to stop once started and may turn into a 'vice'. To interpret such behaviours as a sign of boredom is not to dismiss them as a relatively mild form of incidental suffering. On the contrary, the term boredom signifies the first stage of a form of chronic suffering which threatens the animal's behavioural and subjective integrity.

Boredom may be defined as that stage of behavioural fixation in which the animal's orientation towards a novel stimulus loses its inquisitive and manipulative character. The animal's tendency to act upon the object as a substrate for interaction will decrease and it will investigate the object more hesitantly, with a greater tendency to withdraw. Because the animal's response presumably has not entirely lost its active character at this stage, but is still directed towards the novel object to a certain extent, redirected and stereotyped patterns may initially disappear. However, because a few novel stimuli are not expected to be sufficient to alleviate the animal's state of boredom, such patterns will presumably return after interest in the novel stimuli has waned. If, on the other hand, the animal were to be permanently transferred to an enriched environment, it may be expected that the animal would return to its normally variable and flexible behavioural repertoire.

Depression and/or anxiety

The behaviour of animals which are permanently housed in impoverished conditions may eventually show a high degree of fixation. Stereotyped behaviour patterns become established in the behavioural repertoire, which makes a return to normally flexible behaviour difficult, if not impossible. The animal may spend long periods of time in immobile postures and/or it may respond fearfully or aggressively to unexpected events. In the final stages of behavioural fixation, behavioural diversity is reduced to a minimum and stereotyped behaviour becomes increasingly 'self-directed' (Dantzer, 1986). In schizophrenic and autistic children, the development of self-directed and self-injurious stereotypies has been interpreted as a process of psychological regression in which the child comes to experience its body as an *external* physical stimulus rather than as an intrinsic part of its subjective self (Lourie, 1949; Green, 1985). Likewise, the performance of 'self-directed' stereotypies in captive animals may indicate that the animal has lost its sense of subjective selfhood.

The final stages of behavioural fixation may be described as depression and/or anxiety. Stereotyped and apathetic behaviour may be interpreted as depression (Fox, 1986), while abnormally aggressive and fearful reactions may reflect anxiety (Rowan, 1988). Mostly, however, both states will alternately affect the animal, which may involve considerable fluctuations in arousal. At this stage, the animal no longer experiences a dynamic relationship with its environment, but becomes utterly passive (i.e. in a subjective sense; it may still show considerable motoric activity). To understand that a state of chronic passivity will never be experienced as 'normal' by an animal, even if it has never known any other situation, the concept of psychological time is essential. It implies that if animals cannot maintain the anticipatory character of their behaviour through interaction with the environment, they will inevitably be trapped in physical time. They will experience their environment as monotonous and their own subjective existence as meaningless. However, to apply the concepts of depression and anxiety as used in human psychiatry to animals which are housed in impoverished environments might meet with serious opposition.

In the first place, it may be questioned whether it is justified to use the terms depression and anxiety to designate a state which arises out of and succeeds upon a state of boredom. The concepts of depression and anxiety have been used to describe many different types of human pathology. In animals as well, the term depression has been used in a variety of contexts. It has been related to situations in which animals are confronted with uncontrollable electric shock (the 'learned helplessness' paradigm (Seligman, 1975)), with a dominant animal from which they have previously lost fights (as shown in rats: Koolhaas *et al.*, 1990), or with a mate they do not accept (as in tree shrews: von Holst, 1986). I suggest, however, that such different routes may all lead to a similar end-state of general behavioural impairment. This would imply that the term depression may be used to describe suffering in a variety of contexts, including that which arises out of a state of boredom in impoverished environments.

In the second place, it may be argued that the concepts of depression and anxiety can apply only to beings which are capable of experiencing a subjective self in a way similar to human beings. If the human experience of self is interpreted as an introspective, reflective ability, not many animal species will be accredited with the capacity to suffer from depression. In this chapter, however, I have argued that the capacity for subjective experience is, in animals and humans alike, related to the capacity to interact with the environment. This view is in agreement with theories of anxiety such as that of Gray (1982) who relates depression and anxiety to an impairment of the septohippocampal system and connected neural pathways. In such a model, the postulation of a fundamental difference between human and animal depression and anxiety is unwarranted.

Depression and/or anxiety may then be defined as that stage of behavioural fixation in which the animal has lost its tendency to approach and investigate novel stimuli. Instead, it attempts to avoid such stimuli in one way or another. Accordingly, stereotyped patterns are expected to persist in the behavioural repertoire. Explorative and playful elements may be eliminated to such an extent that the animal's behavioural repertoire loses its versatility. It no longer resembles a species-specific time-budget but is generally disorganized and rigid.

In summary, I suggest that animals experience an environment as 'monotonous', that is as remaining endlessly the same, when the environment does not provide sufficient opportunity for behavioural interaction. 'Monotony' is thus defined as a parameter of the animal–environment relationship, and not as a trait of the physical environment by itself. An environment with a high level of physical change could also be experienced as monotonous by an animal if it is prevented from interacting with that environment.

Monotonous environments deprive animals of their behavioural and subjective integrity. The hypothesis that animals suffer seriously under such conditions involves neither anthropomorphic projection, nor a disrespect for forms of suffering which are uniquely human. What it provides is an acknowledgement of the intensely felt, non-verbal quality of subjective awareness in which animals, children, the mentally handicapped and adult human beings are united.

Summary and Conclusion

In this chapter, I have proposed a model of animal boredom and depression. Evidence is discussed which indicates that the development of stereotyped behaviour reflects the impairment of an animal's capacity to interact with the environment. Behaviour acquires an increasingly rigid and mechanical character; it is determined more and more by immediately available environmental stimuli and loses its innovative, anticipatory nature. It is argued that, because the capacity to interact with the environment signifies the irreducibly subjective nature of behaviour, such a process may be regarded as direct evidence of chronic suffering. The gradual impairment of an animal's subjective integrity may be conceived as boredom in the early stages of impairment and as depression and/or anxiety in later, more severe stages of impairment. These states may be distinguished experimentally by measuring the animal's tendency to actively orient towards novel stimuli.

Acknowledging the dynamic, active character of animal behaviour urges us to provide captive animals with an environment which facilitates the expression of species-specific behaviour. A reasonable amount of environmental complexity and variability will be needed to allow fulfilment of species-specific time-budgets, including the performance of explorative and/or playful behaviours. The animal must be given the chance to set its own goals, and to do the things it wants to do in its own good time.

Acknowledgements

I wish to thank Tjard de Cock Buning, Alistair Lawrence, Diederik van Liere, Jeff Rushen, Henk Verhoog and Cor van der Weele for their highly valuable and clarifying comments on previous versions of this chapter. Work for this chapter was supported financially by grant no. 8701 of the Ministry of Welfare, Health, and Cultural Affairs of the Dutch Government.

References

Barmack, J.E. (1938) The effect of benzedrine sulfate (benzyl methyl carbinamine) upon the report of boredom and other factors. *The Journal of Psychology* 5, 125–133.

Barnett, J.L. and Hemsworth, P.H. (1991) The validity of physiological and behavioural measures of animal welfare. *Applied Animal Behaviour Science* 25, 177–187.

Benus, R.F. (1988) Aggression and coping; differences in behavioural strategies between aggressive and non-aggressive male mice. PhD Thesis, University of Groningen, The Netherlands.

Berkson, G. (1983) Repetitive stereotyped behaviours. *American Journal of Mental Deficiency* 88, 239–246.

Berkson, G., Mason, W. and Saxon, S. (1963) Situation and stimulus effects on stereotyped behavior of chimpanzees. *Journal of Comparative and Physiological Psychology* 56, 786–792.

Berlyne, D.E. (1960) *Conflict, Arousal and Curiosity*. McGraw-Hill, London.

Best, J.B. (1963) Protopsychology. *Scientific American* 208, 55–62.

Bindra, D. (1984) Cognition, its origin and future in psychology. In: Royce, J.R. and Mos, L.P. (eds), *Annals of Theoretical Psychology*. Plenum Press, New York.

Birch, H.G. (1945) The relation of previous experience to insightful problem-solving. *Journal of Comparative Psychology* 38, 367–383.

Birke, L.I.A. and Archer, J. (1983) Some issues and problems in the study of animal exploration. In: Archer, J. and Birke, L.I.A. (eds), *Exploration in Animals and Humans*. Van Nostrand Reinhold Inc., London.

Blokhuis, H.J. (1986) Feather pecking in poultry; its relation with ground pecking. *Applied Animal Behaviour Science* 16, 63–67.

Boice, R. (1981) Capacity and feralization. *Psychological Bulletin* 89, 407–421.

Branch, M.N. (1982) Misrepresenting behaviorism. *Behavioral and Brain Sciences* 5, 372–373.

Broom, D.M. (1986) Stereotypies and responsiveness as welfare indicators in stall-housed sows. *Animal Production* 42, 438–439.

Broom, D.M. (1988) The scientific assessment of animal welfare. *Applied Animal Behaviour Science* 20, 5–19.

Buchenauer, D. (1981) Parameters for assessing welfare, ethological criteria. In: Sybesma, W. (ed.), *The Welfare of Pigs*. Martinus Nijhoff, Dordrecht.

Butler, R.A. and Alexander, H.M. (1955) Daily patterns of visual exploratory behavior

in the monkey. *Journal of Comparative and Physiological Psychology* 48, 247–249.
Butler, R.A. and Harlow, H.F. (1954) Persistence of visual exploration in monkeys. *Journal of Comparative and Physiological Psychology* 47, 258–263.
Butterworth, G. and Hopkins, B. (1988) Hand-mouth coordination in the new-born baby. *British Journal of Developmental Psychology* 6, 303–314.
Buytendijk, F.J.J. (1938) *Wege zum Verständnis der Tiere*. Niehans, Zürich.
Buytendijk, F.J.J. (1953) Toucher et être touché. *Archives Néerlandaise de Zoologie* 10, 34–44.
Catania, A.C. (1982) Antimisrepresentationalism. *The Behavioral and Brain Sciences* 5, 374–375.
Chamove, A.S. and Anderson, J.R. (1981) Self-aggression, stereotypy and self-injurious behaviour in man and monkeys. *Current Psychological Reviews* 1, 245–256.
Cooper, J.J. and Nicol, C.J. (1991) Stereotypic behaviour affects environmental preference in bank voles. *Animal Behaviour* 41, 971–979.
Cronin, G.M. and Wiepkema, P.R. (1984) An analysis of stereotyped behaviour in tethered sows. *Annales de la Recherche Veterinaire* 15, 263–270.
Daanje, A. (1951) On locomotory movements in birds and the intention movements derived from them. *Behaviour* 3, 48–98.
Dantzer, R. (1986) Behavioural, physiological and functional aspects of stereotyped behavior: a review and a re-interpretation. *Journal of Animal Science* 62, 1776–1786.
Davies, P. and Gribbin, J. (1991) *The Matter Myth, Towards 21st-Century Science*. Viking, London.
Dawkins, M.S. (1980) *Animal Suffering. The Science of Animal Welfare*. Chapman and Hall, London.
Dawkins, M.S. (1990) From an animal's point of view: motivation, fitness and animal welfare. *The Behavioral and Brain Sciences* 13, 1–61.
Dember, W.N., Earl, R.W. and Paradise, N. (1957) Response by rats to differential stimulus complexity. *Journal of Comparative and Physiological Psychology* 50, 515–518.
Dickinson, A. (1985) Actions and habits: the development of behavioural autonomy. *Philosophical Transactions of the Royal Society of London Series B* 308, 67–78.
Duncan, I.J.H. (1970) Frustration in the fowl. In: Freedman, B.M. and Gordon, R.F. (eds), *Aspects of Poultry Behaviour, British Egg Marketing Board Symposium 6*. British Poultry Science, Edinburgh.
Duncan, I.J.H. and Wood-Gush, D.G.M. (1972) Thwarting of feeding behaviour in the domestic fowl. *Animal Behaviour* 20, 444–451.
Einon, D.F. (1983) Play and exploration. In: Archer, J. and Birke, L.I.A. (eds), *Exploration in Animals and Humans*. Van Nostrand Reinhold Inc, London.
Einon, D. and Morgan, M. (1976) Habituation of object contact in socially-reared and isolated rats (*Rattus norvegicus*). *Animal Behaviour* 24, 415–420.
Einon, D.F., Morgan, M.J. and Kibbler, C.C. (1978) Brief periods of socialization and later behavior in the rat. *Developmental Psychobiology* 11, 213–225.
Epstein, R. (1982) Representation: A concept that fills no gaps. *The Behavioral and Brain Sciences* 5, 377–378.
Fagen, R. (1982) Evolutionary issues in development of behavioral flexibility. In:

Bateson, P.P.G. and Klopfer, P.H. (eds), *Perspectives in Ethology, Vol. 5*. Plenum Press, New York.

Fagen, R. (1984) Play and behavioural flexibility. In: Smith, P.K. (ed.), *Play in Animals and Humans*. Basil Blackwell, London.

Fentress, J.C. (1976) Dynamic boundaries of patterned behaviour: interaction and self-organization. In: Bateson, P.P.G. and Hinde, R.A. (eds), *Growing Points in Ethology*. Cambridge University Press, Cambridge.

Ferchmin, P.A. and Eterovic, V.A. (1977) Brain plasticity and environmental complexity: role of motor skills. *Physiology and Behavior* 18, 455–461.

Ferchmin, P.A., Bennett, E.L. and Rosenzweig, M.R. (1975) Direct contact with enriched environment is required to alter cerebral weights in rats. *Journal of Comparative and Physiological Psychology* 88, 360–367.

Ferchmin, P.A., Eterovic, V.A. and Levin, L.E. (1980) Genetic learning deficiency does not hinder environment-dependent brain growth. *Physiology and Behavior* 24, 45–50.

Fox, M.W. (1986) *Laboratory Animal Husbandry, Ethology, Welfare and Experimental Variables*. State University of New York Press, Albany.

Gallistel, C.R. (1990) *The Organization of Learning*. MIT Press, Cambridge, Massachusetts.

Gardner, R.A. and Gardner, B.T. (1988) Feedforward versus feedbackward: An ethological alternative to the law of effect. *The Behavioral and Brain Sciences* 11, 429–493.

Goss, A.E. and Wischner, G.J. (1956) Vicarious trial and error and related behaviour. *Psychological Bulletin* 53, 35–54.

Graumann C.F. and Sommer, M. (1984) Perspectives on cognitivism. In: Royce, J.R. and Mos, L.P. (eds), *Annals of Theoretical Psychology*. Plenum Press, New York.

Gray, J.A. (1982) Précis of 'The neurophysiology of anxiety: an enquiry into the functions of the septo-hippocampal system'. *The Behavioral and Brain Sciences* 5, 469–534.

Green, A.H. (1985) Self-mutilation in schizophrenic children. Reprinted in: Murphy, G. and Wilson, B. (eds), *Self-injurious Behaviour*. British Institute of Mental Handicap, Kidderminster.

Griffin, D.R. (1976) *The Question of Animal Awareness*. The Rockefeller University Press, New York.

Griffin, D.R. (1984) *Animal Thinking*. Harvard University Press, Cambridge, MA.

Groves, P.M. and Thompson, R.F. (1970) Habituation: a dual process theory. *Psychological Review* 77, 419–450.

Harlow, H.R. (1950) Learning and satiation of response in intrinsically motivated complex puzzle performance in monkeys. *The Journal of Comparative and Physiological Psychology* 43, 289–294.

Harlow, H.R., Harlow, M.K. and Meyer, D.R. (1950) Learning motivated by a manipulation drive. *Journal of Experimental Psychology* 40, 228–234.

Hawking, S. (1988) *A Brief History of Time*. Bantam, New York.

Heidegger, M. (1927) *Sein und Zeit*. I. Niemeyer, Halle.

Held, R. (1961) Exposure-history as a factor in maintaining stability of perception and coordination. *Journal of Nervous and Mental Disease* 132, 26–32.

Held, R. and Freedman, S.J. (1963) Plasticity in human sensorimotor control. *Science* 142, 455–462.

Held, R. and Hein, A. (1963) Movement-produced stimulation in the development of visually guided behaviour. *Journal of Comparative and Physiological Psychology* 56, 872–876.

Hinde, R.A. (1970). *Animal Behaviour*. McGraw-Hill, London.

Holst, D. von (1986) Vegetative and somatic components of tree shrews' behavior. *Journal of the Autonomic Nervous System, (Suppl.)* 657–670.

Holst, D. von and Kolb, H. (1976) Sniffing frequency of *Tupaia belangeri*: a measure of central nervous activity (arousal). *Journal of Comparative Physiology* 105, 243–257.

Holst, D. von and Mittelstaedt, H. (1950) Das Reafferenzprinzip. *Naturwissenschaften* 37, 464–476.

Hughes, B.O. and Duncan, I.J.H. (1988) The notion of ethological 'need', models of motivation and animal welfare. *Animal Behaviour* 36, 1696–1707.

Hughes, M. (1983) Exploration and play in young children. In: Archer, J. and Birke, L.I.A. (eds), *Exploration in Animals and Humans*. Van Nostrand Reinhold Inc., London.

Hutt, C. (1966) Exploration and play in children. *Symposium of the Zoological Society of London* 18, 61–81.

Hutt, C. (1970) Specific and diverse exploration. In: Reese, H.W. and Lipsitt, L.P. (eds), *Advances in Child Development and Behavior 5*. Academic Press, New York.

Hutt, C. and Hutt, S.J. (1965) The effects of environmental complexity on stereotyped behaviour of children. *Animal Behaviour* 13, 1–4.

Inglis, I.R. (1983) Towards a cognitive theory of exploratory behaviour. In: Archer, J. and Birke, L.I.A. (eds), *Exploration in Animals and Humans*. Van Nostrand Reinhold Inc, London.

Jackson, T.A. (1942) Use of the stick as a tool by young chimpanzees. *Journal of Comparative Psychology* 34, 223–235.

Kahneman, D. (1973) *Attention and Effort*. Prentice Hall, Englewood Cliffs, New Jersey.

Kastelein, R.A. and Wiepkema, P.R. (1989) A digging trough as occupational therapy for Pacific walruses (*Odobenus rosmarus divergens*). *Aquatic Mammals* 15, 9–17.

Keiper, R.R. (1970) Studies of stereotypy function in the canary (*Serinus canarius*). *Animal Behaviour* 18, 353–357.

Kish, G.B. (1955) Learning when the onset of illumination is used as reinforcing stimulus. *Journal of Comparative Psychology* 48, 261–264.

Koegel, R.L., Firestone, P.B., Kramme, K.W. and Dunlap, G. (1974) Increasing spontaneous play by suppressing self-stimulation in autistic children. *Journal of Applied Behavior Analysis* 7, 521–528.

Komisaruk, B.R. (1970) Synchrony between limbic system theta activity and rhythmical behavior in rats. *Journal of Comparative and Physiological Psychology* 70, 483–492.

Koolhaas, J.M., Hermann, P.M., Kemperman, C., Bohus, B., Hoofdakker, R.H. van den and Beersma, D.G.M. (1990) Single social defeat in male rats induces a gradual but long-lasting behavioural change: a model of depression? *Neuroscience Research Communications* 7, 35–41.

Krechevsky, I. (1932) 'Hypothesis' versus 'chance' in the pre-solution period in sensory discrimination learning. Reprinted in: Levine, M. (ed.), *A Cognitive Theory of Learning*. Lawrence Erlbaum Associates, Hillsdale, 1985.

Krechevsky, I. (1936) Brain mechanisms and brightness discrimination learning. *Journal of Comparative Psychology* 21, 405–441.

Kuijper, S. (1963) Einige aspecten van vrij en dwangmatig herhalen (Some aspects of free and compulsive repetitive behaviour). PhD Thesis, University of Groningen, The Netherlands.

Kummer, H. and Goodall, J. (1985) Conditions of innovative behaviour in primates. *Philosophical Transactions of the Royal Society of London Series B* 308, 203–214.

Levine, M. (1959) A model of hypothesis behaviour in discrimination learning set. *Psychological Review* 66, 353–366.

Libet, B. (1985) Unconscious cerebral initiative and the role of conscious will in voluntary action. *The Behavioral and Brain Sciences* 8, 529–566.

Lourie, R.S. (1949) The role of rhythmic patterns in childhood. *American Journal of Psychiatry* 105, 653–660.

Lovaas, O.I., Litrownik, A. and Mann, R. (1971) Response latencies to auditory stimuli in autistic children engaged in self-stimulatory behavior. *Behavioral Research and Therapy* 9, 39–49.

Maier, N.R.F. (1949) *Frustration. The Study of Behaviour without a Goal*. McGraw-Hill, New York.

Martin, P. and Caro, T.M. (1985) On the functions of play and its role in behavioral development. *Advances in the Study of Behavior* 15, 59–103.

Mason, G.J. (1991) Stereotypies and suffering. *Behavioural Processes* 25, 103–115.

Mason, W.A. and Green, P.C. (1962) The effects of social restriction on the behavior of rhesus monkeys: IV. Responses to a novel environment and to an alien species. *Journal of Comparative and Physiological Psychology* 55, 363–368.

McBride, G. (1980) Adaptation and welfare at the man-animal interface. In: Wodzicka, M., Edey, T.N. and Lynch, J.J. (eds), *Reviews in Rural Science IV*, University of New England, Armidale, New South Wales.

McFarland, D.J. (1989) *Problems of Animal Behaviour*. Longman Scientific and Technical, Harlow.

McKinnon, A.J., Edwards, S.A., Stephens, D.B. and Walters, D.E. (1989) Behaviour of groups of weaner pigs in three different housing systems. *British Veterinary Journal* 145, 367–372.

Merleau-Ponti, M. (1942) *La Structure du Comportement*. PUF, Paris.

Meyer-Holzapfel, M. (1968) Abnormal behaviour in zoo animals. In: Fox, M.W. (ed.) *Abnormal Behaviour in Animals*. W.B. Saunders, Philadelphia.

Militzer, K. and Wecker, E. (1986) Behaviour-associated alopecia areata in mice. *Laboratory Animals* 20, 9–13.

Morgan, M.J. (1973) Effects of post-weaning environment on learning in the rat. *Animal Behaviour* 21, 429–442.

Morgan, M.J., Einon, D.F. and Nicholas, D. (1975) Effects of isolation rearing on behavioural inhibition in the rat. *Quarterly Journal of Experimental Psychology* 27, 615–634.

Morris, D. (1964) The response of animals to a restricted environment. *Symposium of the Zoological Society of London* 13, 99–118.

Mowrer, O.H. (1960) *Learning Theory and the Symbolic Processes*. Wiley, New York.

Mowrer, O.H. and Ullman, A.D. (1945) Time as a determinant in integrative learning. *Psychological Review* 52, 61–90.

Nagel, T. (1974) What is it like to be a bat? *Philosophical Review* 83, 435–451.

Ödberg, F.O. (1987) The influence of cage size and environmental enrichment on the development of stereotypies in bank voles (*Clethrionomys glareolus*). *Behavioural Processes* 14, 155–173.

O'Keefe, J. and Nadel, L. (1978) *The Hippocampus as a Cognitive Map*. Clarendon Press, Oxford.

Posner, M.I. (1978) *Chronometric Explorations of the Mind*. Lawrence Erlbaum Associates, Hillsdale, New Jersey.

Pribram, K.H. and McGuinness, D. (1975) Arousal, activation, and effort in the control of attention. *Psychological Review* 82, 116–149.

Putten, G. van and Dammers, J. (1976) A comparative study of the well-being of piglets reared conventionally and in cages. *Applied Animal Ethology* 2, 339–356.

Renner, M.J. and Rosenzweig, M.R. (1986) Object interactions in juvenile rats (*Rattus norvegicus*): effects of different experiential histories. *Journal of Comparative Psychology* 100, 229–236.

Rohrbaugh, J.W. (1984) The orienting reflex: performance and central nervous system manifestations. In: Parasuraman, R. and Davies, D.R. (eds), *Varieties of Attention*. Academic Press Inc., Orlando.

Roitblat, H.L. (1982) The meaning of representation in animal memory. *The Behavioral and Brain Sciences* 5, 353–406.

Rowan, A.N. (1988) Animal anxiety and animal suffering. *Applied Animal Behaviour Science* 20, 135–142.

Rushen, J. (1985) Stereotypies, aggression, and the feeding schedules of tethered sows. *Applied Animal Behaviour Science* 14, 137–147.

Russell, P.A. (1983) Psychological studies of exploration in animals: a reappraisal. In: Archer, J. and Birke, L.I.A. (eds), *Exploration in Animals and Humans*. Van Nostrand Reinhold Inc, London.

Ryle, G. (1949) *The Concept of Mind*. Hutchinson's University Library, London.

Sachser, N. and Lick, C. (1991) Social experience, behavior and stress in guinea-pigs. *Physiology and Behavior* 50, 1–8.

Schouten, W.G.P. and Wiepkema, P.R. (1991) Coping styles of tethered sows. *Behavioural Processes* 25, 125–132.

Seligman, M.E.P. (1975) *Helplessness: On Depression, Development and Death*. W.H. Freeman, San Francisco.

Sevenster, P. and Roosmalen, M. van (1985) Cognition in sticklebacks: some experiments on operant conditioning. *Behaviour* 93, 170–183.

Skinner, B.F. (1984a) Methods and theories in the experimental analysis of behavior. *The Behavioral and Brain Sciences* 7, 511–546.

Skinner, B.F. (1984b) Behaviorism at fifty. *The Behavioral and Brain Sciences* 7, 615–621.

Smith, P.K. and Dutton, S. (1979) Play and training in direct and innovative problem solving. *Child Development* 50, 830–836.

Stahl, J.M., O'Brien, R.A. and Hanford, P.V. (1973) Visual exploratory behavior in the pigeon. *Bulletin of the Psychonomic Society* 1, 35–36.

Stevenson, M.F. (1983) The captive environment: its effect on exploratory and related behavioural responses in wild animals. In Archer, J. and Birke, L.I.A. (eds), *Exploration in Animals and Humans*. Van Nostrand Reinhold Inc, London.

Stolba, A. and Wood-Gush, D.G.M. (1989) The behaviour of pigs in a semi-natural environment. *Animal Production* 48, 419–425.

Stolba, A., Baker, N. and Wood-Gush, D.G.M. (1983) The characterisation of stereotyped behaviour in stalled sows by informational redundancy. *Behaviour* 87, 157–182.

Sylva, K., Bruner, J. and Genova, P. (1974) The role of play in the problem-solving of children 3–5 years old. In: Bruner, J., Jolly, A. and Sylva, K. (eds), *Play*. Penguin, London.

Terlouw, E.M.C., Lawrence, A.B. and Illius, A.W. (1991) Influences of feeding level and physical restriction on development of stereotypies in sows. *Animal Behaviour* 42, 981–993.

Thelen, E. (1979) Rhythmical stereotypes in normal human infants. *Animal Behaviour* 27, 699–715.

Thelen, E. (1981a) Kicking, rocking, and waving: contextual analysis of rhythmical stereotypies in normal human infants. *Animal Behaviour* 29, 3–11.

Thelen, E. (1981b) Rhythmical behavior in infancy: an ethological perspective. *Developmental Psychology* 17, 237–257.

Thinus-Blanc, C. (1988) Animal spatial cognition. In: Weiskrantz, L. (ed.), *Thought without Language*. Clarendon Press, Oxford.

Thompson, R.F., Berry, S.D., Rinaldi, P.C. and Berger, T.W. (1979) Habituation and the orienting reflex: The dual-process theory revisited. In: Kimmel, H.D., van Olst, E.H. and Orlebeke, J.F. (eds), *The Orienting Reflex in Humans*. Lawrence Erlbaum Associates, Hillsdale.

Tolman, E.C. (1932) *Purposive Behaviour in Animals and Man*. Appleton-Century, New York.

Tolman, E.C. (1948) Cognitive maps in rats and men. *Psychological Review* 55, 189–208.

Vastrade, F.M. (1986) The social behaviour of free-ranging domestic rabbits (*Oryctolagus cuniculus* L.). *Applied Animal Behaviour Science* 16, 165–177.

Wechsler, B. (1991) Stereotypies in polar bears. *Zoo Biology* 10, 177–188.

Weizsäcker, V. von (1967) *Der Gestaltkreis*, 4th edition. Stuttgart.

Welker, W.I. (1956) Some determinants of play and exploration in chimpanzees. *Journal of Comparative and Physiological Psychology* 49, 84–89.

Welker, W.I. (1964) Analysis of sniffing of the albino rat. *Behaviour* 22, 223–244.

Wemelsfelder, F. (1990) Boredom and laboratory animal welfare. In: Rollin, B.E. (ed.), *The Experimental Animal in Biomedical Research*. CRC Press, Boca Raton.

White, R.W. (1959) Motivation reconsidered: the concept of competence. *Psychological Review* 66, 297–333.

Wiepkema, P.R. (1987) Behavioural aspects of stress. In: Wiepkema, P.R. and Adrichem, P.W.M. (eds), *Biology of Stress in Farm Animals: An Integrative Approach*. Martinus Nijhoff Publishers, Dordrecht.

Wood-Gush, D.G.M. (1972) Strain differences in response to sub-optimal stimuli. *Animal Behaviour* 20, 72–76.

Wood-Gush, D.G.M. and Vestergaard, K. (1991) The seeking of novelty and its relation to play. *Animal Behaviour* 42, 599–606.

Wood-Gush, D.G.M., Duncan, I.J.H. and Fraser, D. (1975) Social stress and welfare problems in agricultural animals. In: Hafez, E.S.E. (ed.), *The Behaviour of Domestic Animals*, 3rd edition. Ballière Tindall, London.

Wood-Gush, D.G.M., Stolba, A. and Miller, C. (1983) Exploration in farm animals and animal husbandry. In: Archer, J. and Birke, L.I.A. (eds), *Exploration in Animals and Humans*. Van Nostrand Reinhold Inc., London.

Wood-Gush, D.G.M., Vestergaard, K. and Petersen, V. (1990) The significance of motivation and environment in the development of exploration in pigs. *Biology of Behaviour* 15, 39–52.

Zelazo, P.R. and Kearsley, R.B. (1980) The emergence of functional play in infants: evidence for a major cognitive transition. *Journal of Applied Developmental Psychology* 1, 95–117.

Stress and the Physiological Correlates of Stereotypic Behaviour 5

JAN LADEWIG[1], ANNE MARIE DE PASSILLÉ[2], JEFFREY RUSHEN[2], WILLEM SCHOUTEN[3], E.M. CLAUDIA TERLOUW[4] AND EBERHARD VON BORELL[5]

[1]*Institut fur Tierzucht und Tierverhalten, Trenthorst, Germany:* [2]*Agriculture Canada Research Station, Lennoxville, Quebec, Canada:* [3]*Wageningen Agricultural University, The Netherlands:* [4]*The Scottish Agricultural College, Edinburgh, UK:* [5]*Iowa State University, Ames, Iowa, USA.*

Editors' Introductory Notes:
It is generally assumed that stereotypies arise in response to stress, and it is not surprising therefore that an increasing number of studies have investigated the relationship between stereotypies and conventional physiological measures of stress.

As an introduction to their review of studies of stress and stereotypies in large animals, Ladewig *et al.* consider the development of the stress concept itself. Ladewig *et al.* see stress as a process involving the interaction between external events ('stressors') and individual predisposition(s) (determined by genetic factors and early experience), that gives rise to measurable 'stress' responses. This approach makes a number of useful points, such as emphasizing the 'post-Selye' move towards more consideration of the psychological aspects of the stress response. For example, although stressors can have measurable physical properties, it is their psychogenic properties (e.g. contextual cues surrounding exposure to the stressor) that will determine their effect on the animal. The emphasis on the individual nature of the stress response is also welcome in the light of empirical evidence on the large inter-individual variability in stress responses. Last, the animal can respond to stressors with a variety of physiological systems; Ladewig *et al.*, highlight the role of the hypothalamus and in particular corticotrophic releasing hormone (CRH) in coordination of the physiological correlates of stress, but other neurochemical and physiological systems should not be overlooked.

In the second part of the chapter, various physiological systems are considered with respect to their relationships with stereotypies; the emphasis here is on those systems (endogenous opioid peptides; hypothalamic-pituitary-axis (HPA); autonomic nervous system (ANS); reproductive and digestive hormones) that have received substantial attention in large (rather than laboratory) animals. The broad conclusion from this review is of a lack of consistent relationships between stereotypies and the various physiological correlates. For example, although opioids have been found to inhibit performance of stereotypies their effectiveness in doing so varies substantially between studies; one study actually reported an increase in performance of an apparently stereotypic activity. Even greater contradictions are found when the HPA is considered. While some studies report negative correlations between performance of stereotypies and circulating glucocorticoids, others have reported little evidence of such an effect.

Ladewig et al. suggest that in general the heterogeneity of stereotypies and the complexities of the neuroendocrine systems involved underlie the failure to demonstrate a relationship between stress and stereotypies. This emphasizes the need for greater attention to be given to the age of the stereotypy and the past history of the subject (see Chapter 2). Ladewig et al. also suggest that a better understanding of stress and stereotypies will result if attention is focused on measurements of brain neuroendocrine activity (particularly the role of CRH), rather than just on peripheral blood measures.

Introduction

The importance of physiological measurements in stress research on domestic animals has often been emphasized (e.g. Dantzer and Mormède, 1983a; Smidt, 1983; Ladewig and von Borell, 1988). In the study of stereotypic behaviour, this importance is particularly obvious since the 'function' of this type of behaviour is difficult to explain from a behavioural standpoint alone. As a consequence, the analysis of various hormones has been included in an increasing number of behavioural studies in an attempt to help interpretation of behavioural results. As pointed out by Rushen (1991), however, the results are often contradictory, without any consensus between different studies. The main reasons for these difficulties are due to the complexity of the underlying physiological and psychological mechanisms and to their sensitivity to various intervening factors. Many of these intervening factors cannot be controlled, causing a considerable amount of variation in the results and making a comparison between different studies or replications of studies difficult.

In this chapter, a brief description of the development of the current stress concept will be given, followed by an overview of some of the most important sources of variability. Thereafter, a review of physiological correlates of environmental stereotypies is given as well as the involvement of the autonomic nervous system and reproductive and gastrointestinal functions in stress and stereotypic behaviour. Particular attention is paid to large animals as the relationship between stress and stereotypic behaviour in rodents is discussed elsewhere (Chapters 6 and 7).

The Concept of Stress

Throughout this century, stress has been the subject of intense investigation. In the beginning, the positive aspects of stress reactions were emphasized. The increased release of some hormones clearly increases the chance of survival of the organism, a fact that nowadays, unfortunately, is often forgotten. Beginning with Cannon's 'Fight and Flight' reaction, which emphasized the role of adrenalin from the adrenal medulla (Cannon, 1929), and Selye's 'General Adaptation Syndrome', which emphasized the role of the glucocorticosteroids from the adrenal cortex (Selye, 1956), attention has gradually shifted towards the psychological involvement in stress reactions. This is due particularly to Mason's 'Cognitive Mediator Concept' (Mason, 1974) and Henry and Stephens' 'Coping-Predictability Concept' (Henry and Stephens, 1977). Furthermore, as most clearly demonstrated by Mason's work, stress reactions do not consist only of activation of the sympatho-adrenomedullary and the hypothalamo-pituitary adrenocortical systems but also of numerous other nueoroendocrine systems that react in an organized and integrated fashion (Mason, 1974). The discovery of the opioid peptides and their involvement in stress reactions in the mid-1970s (Akil *et al.*, 1984) has added a new dimension to the complexity and diversity of the organism's reaction to stress (Grossman, 1988).

Because of this complexity, the present concept of stress is best described by the so-called hour-glass model (Veith-Flanigan and Sandman, 1985) which simply says that 'stress encompasses all possible extra-individual events capable of evoking a broad spectrum of intra-individual responses mediated by a complex filter labelled individual differences'. The advantage of this concept is that it allows a more detailed analysis of the three aspects of stress, namely the stressors, the individual differences, and the stress reactions.

Stressors

In most studies too little attention is paid to the 'extra-individual events' supposed to act as stressors and, particularly, to the context in which these stimuli are applied. The result is that minor differences in or changes to the stressor can dramatically change the stress reaction. As indicated in Fig. 5.1, stressors can be characterized in a qualitative as well as a quantitative way. Qualitatively, it is possible to identify physical stressors (thermal, chemical, electrical, etc.) or stressors defined by the situation (e.g. immobilization, social isolation, etc.). Such a description of the stressor, however, is of only limited value. First, in many cases several stressors are applied simultaneously (e.g. electroshock and social isolation). Second, stressors that differ qualitatively may evoke the same emotional reactions and thus give rise to the same physiological response. And third, in some cases the actual stressor applied

```
Stressor ─┬─ Quality ──┬─ Physical quality (thermal
          │            │   chemical, electrical, etc.)
          │            └─ According to the situation (immobilization,
          │               social isolation, water immersion, etc.)
          └─ Quantity ─┬─ Intensity (degree, decibel, ampere, etc.)
                       └─ Duration ─┬─ Acute (minutes, hours)
                                    ├─ Chronic (days, weeks, months)
                                    └─ Chronic intermittent
```

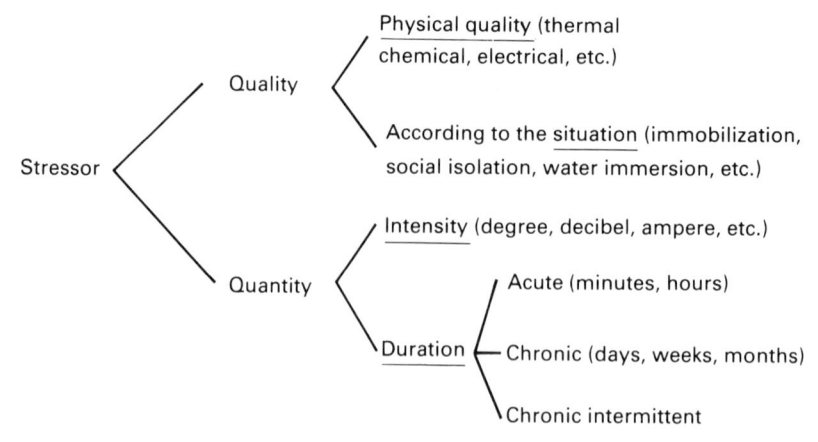

Fig. 5.1. Diagram of possible stressors.

may be less important than the context in which it is applied. Thus, a young goat will respond considerably more to stress if socially isolated than if it is kept in the company of its dam (Lyons et al., 1988).

It is possible to quantify stressors by their intensity and their duration. Although the intensity of some stressors can be measured in physical terms (e.g. heat stress by temperature, electroshocks by voltage and ampere), this measurement is not indicative of the subjective experience of the animal. The same is true for situational stressors in that it is not possible to directly compare the intensity of restraint stress with that of water immersion, for example.

As far as the duration of a stressor is concerned, it is common to distinguish between acute and chronic type stressors; acute stressors usually last minutes or hours and chronic stressors last days, weeks or months. Considering the importance of cognitive involvement in stress, however, it is questionable whether true chronic stress is as common as generally believed. For instance, cattle kept on concrete floors often show disturbed lying-down behaviour (Ladewig and Smidt, 1989; Müllet et al., 1989), as if they have problems bringing their body down into a lying position on the hard surface. During periods of standing and during periods of lying their behavioural and physiological reactions to the hard floor appear less pronounced than while they are lying down. Therefore, although the concrete floor is constantly present, its negative effect is exerted with varying intensity. Since most everyday stressors fluctuate in intensity, a third type has been suggested: chronic intermittent stress (Burchfield, 1979). This form of stressor consists of an acute type stressor that is repeated with a regular or irregular interval.

The reaction of the animal to repeated stressors depends on a number of factors, such as the intensity of the stressor, the duration of each application, the total number of repetitions, and the frequency of the application (also called the interstressor interval; Natelson et al., 1988). With less intense stressors (e.g. a painless restraint of limited duration), the stress reaction decreases with each repetition until no reaction is discernible. The same laws that apply to habituation seem to be in operation: the less intense the stressor, the more frequent its repetition, and the shorter the interstressor interval, the faster the adaptation to the stressor occurs (Natelson et al., 1988; De Boer et al., 1990; Pitman et al., 1990). With more intense stressors (e.g. stressors that remain painful independently of how often they are repeated), either no change in the stress reaction occurs or an increasingly intense reaction follows, a process referred to as sensitization (Konarska et al., 1990; Pitman et al., 1990; see also Chapters 6 and 7).

An open question regarding this type of stressor is whether the changes (i.e. adaptation or sensitization) are due to central nervous processes (possibly on the psychological level, or the animal's 'subjective assessment of the stressor') or to biochemical changes in some peripheral organ, such as changes in the biosynthesis, storage, release, or uptake of a hormone, or receptor modulation (De Boer et al., 1988). Possibly, with less intense stressors (e.g. stressors that are aversive due to their novelty) adaptation at the psychological level is more likely, whereas more intense stressors are more likely to induce biochemical changes. Ladewig and Smidt (1989) have shown that adaptation of the pituitary–adrenocortical axis of cattle to tethering is partly due to a reduction in the sensitivity of the adrenal cortex. Furthermore, it is possible that different hormone systems are affected differently during the same type of chronic intermittent stress and that while some hormones show adaptation, others show no change or sensitization (Konarska et al., 1990).

In any type of stress research, it is necessary to define all stressors involved as precisely as possible, a fact that also applies to systematic investigations on stereotypic behaviour. It is very likely that different situations that induce (the same or different types of) stereotypic behaviour also affect the central nervous processes and the resultant biochemical reactions differently. Unfortunately, stress is often discussed in studies of stereotypic behaviour without the actual stressor being characterized.

Individual Differences

In all behavioural research a dominating source of variability stems from individual differences. Apart from species-specific differences, individual differences are due partly to the genetic predisposition of the animal and partly to its earlier experience, pre- as well as postnatal. This variability is particularly obvious in animals subjected to stress, such as in the situations used to induce stereotypic behaviour. The usual solution to this problem is to increase the number of experimental subjects, but this is not always possible in physiological investigations. Therefore, other methods must be used to reduce variability. Apart from using animals that are genetically related and reared under identical conditions, it is possible to screen subjects prior to an experiment and to establish subgroups according to some criterion (e.g. behavioural response to a 'standard stressor' see Benus, 1988; Lawrence et al., 1991). Physiologically, this approach has been used in relation to the adrenal function test which is sometimes used to diagnose chronic stress. The test consists of stimulation of the adrenal cortex, either with a standard stressor or a standard dosage of synthetic ACTH, and measurement of the resulting corticosteroid release. Numerous studies have shown that chronic stress exerts either an enhancing (e.g. von Borell and Ladewig, 1989) or reducing (e.g. Ladewig and Smidt, 1989) effect on the adrenocortical response pattern. Because this response pattern shows large individual variation and seems to be a constant individual characteristic of the animals (Hennessy et al., 1988; von Borell and Ladewig, 1989), it is possible to screen the experimental subjects prior to a study, to establish subgroups according to the response pattern and, thus, to examine the environmental effect in greater detail (von Borell and Ladewig, 1989, 1992).

In systematic studies on the development of stereotypic behaviour, it is often found that stereotypies develop with different intensity ranging from no stereotypic behaviour at all to intense behaviour, with all shades of intensity in between (e.g. Terlouw et al., 1991). Although the subjects showing the behaviour seem to be the most interesting, it is important to include non-responders as well, not only as control animals for comparison with responders, but also because the neuroendocrine processes in these animals may supply as important information on the mechanisms involved in stereotypic behaviour.

Stress Reactions

Mason's 'Cognitive Mediator Concept' and Henry and Stephens' 'Coping-Predictability Concept' both emphasize the involvement of the central nervous system (CNS) in stress reactions. In this, the hypothalamus plays an essential role, constituting the 'interface' between the CNS and the various endocrine systems (Fig. 5.2).

Phylogenetically, the hypothalamus is the oldest part of the brain, the foremost extension of the spinal cord. Incoming information from the body and from the environment via the sensory organs is integrated here and transformed into signals that are sent to other parts of the body. The hypothalamus is the 'thermostat' of the body for regulating body temperature and is involved in many other body functions, such as the regulation of food and water intake, sexual activity, etc. Around the hypothalamus, the limbic system is located, the second oldest part of the brain and the centre of emotions. The amygdala of the limbic system plays a key role not only in aggression but also in motivation in that it enables the animal to distinguish

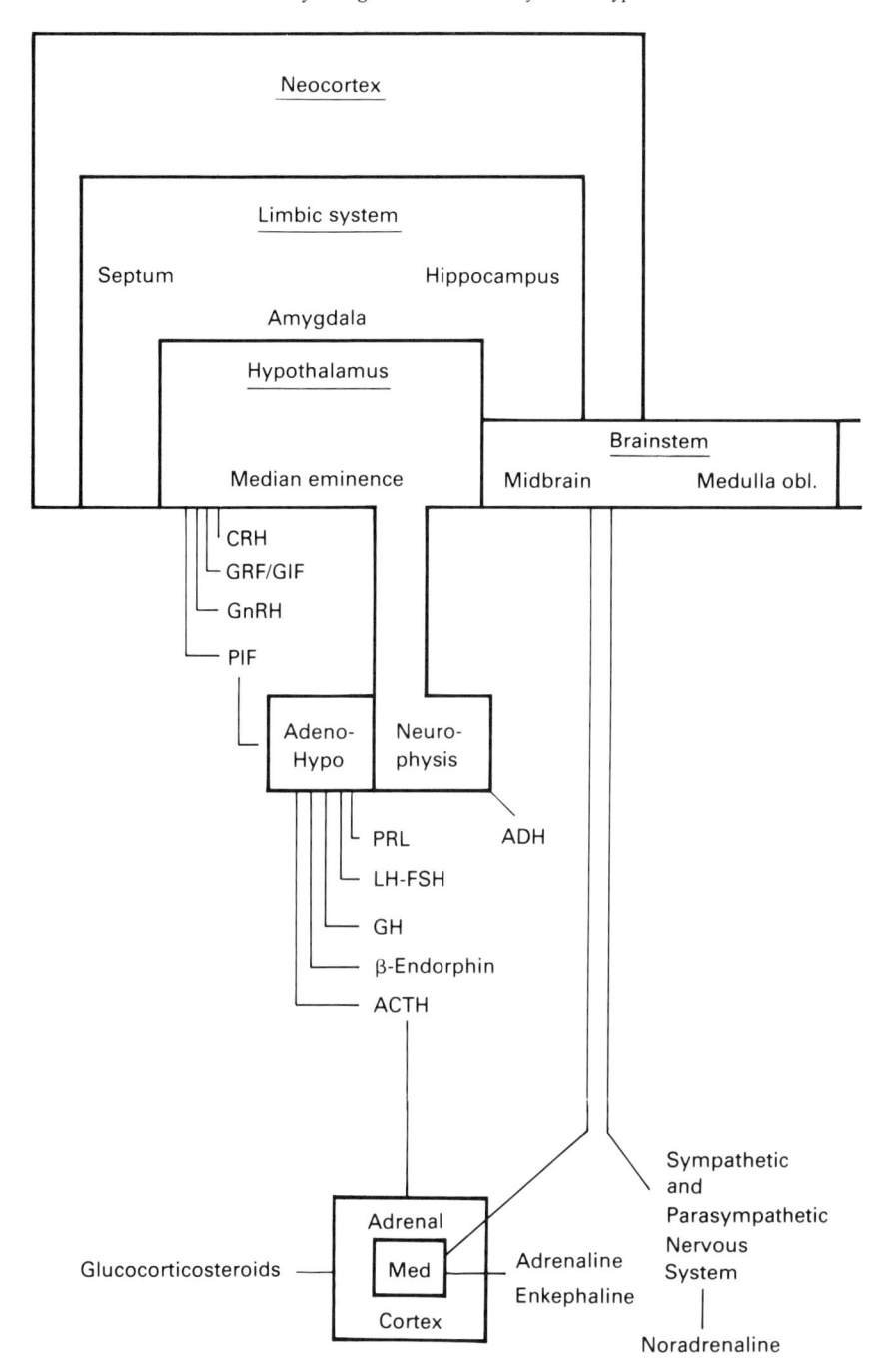

Stress and the Physiological Correlates of Stereotypic Behaviour 103

Fig. 5.2. Diagram of the central nervous–endocrine connections.

between positive stimuli (such as rewards) and negative stimuli (such as aversion). In the hippocampus of the limbic system a comparison is made between the expectations of the animal and its actual situation, both as far as its position in space is concerned and as far as its personal position (such as social status) is concerned. Around the limbic system the youngest part of the brain is located, namely the neocortex, in which particularly cognitive and associative thinking take place.

Information from these higher brain structures as well as from the body reaches the hypothalamus via numerous afferent pathways, and signals are transmitted from the hypothalamus via different efferent pathways. Three types of pathways involved in stress responses have been identified (Fig. 5.2). One pathway – the sympathetic part of the autonomic nervous system – runs through the brain stem and the intermediate horn of the grey matter of the thoracic and lumbar regions of the spinal cord and is responsible for the secretion of the catecholamines adrenaline and noradrenaline. A second pathway consists of neurons of the hypothalamus that send their axons to the neurohypophysis from where the hormones vasopressin and oxytocin are released into the periphery. The third pathway consists of the secretion of releasing or inhibiting factors from the median eminence of the hypothalamus, the transport of these factors via the portal blood to the adenohypophysis, and their stimulation there of a specific pituitary hormone. One example of this efferent pathway is the hypothalamo–pituitary–adrenocortical axis, involving the secretion of corticotropic releasing hormone (CRH) from the median eminence, stimulating secretion of ACTH from the adenohypophysis. The latter hormone stimulates secretion of glucocorticosteroids from the adrenal cortex.

In reality, the chain of events is not as simple as indicated in Fig. 5.2. Although CRH is the most important stimulator of ACTH, other factors, such as vasopressin and catecholamines, exert important effects, too (e.g. Axelrod and Reisine, 1984). Furthermore, stimulation of ACTH is not the only function of CRH. Evidence is slowly accumulating that many reactions typical of stress, such as catecholamine release, inhibition of LH and growth hormone, anxiety, reduced exploratory behaviour, are directly mediated by CRH. The secretion of brain CRH seems to be an essential part of the reaction to stress (Dunn and Berridge, 1990).

The involvement of endogenous opioid peptides in stress reactions has added an enormous degree of complexity to our understanding of neuroendocrinology. Not only do the opioid peptides consist of numerous peptides that exert their effect over many different receptors, but they also act both as typical neurotransmitters and as typical hormones. As neurotransmitters they are able to modulate the transmission of nerve impulses, for example to reduce the sensation of pain (Amit and Galina, 1986; Lewis *et al.*, 1987). As hormones they can interfere with the release or uptake of other hormones (Amir *et al.*, 1980; Székely and Rónai, 1982), especially those involved in reproduction

(Rivier and Rivest, 1991). The close connection between stress and opioids is shown by the opioid-based reduction in pain sensitivity that occurs when animals are stressed (Amit and Galina, 1986) and further supported by the fact that β-endorphin is co-released with ACTH, enkephalin co-released with adrenaline, and dynorphin stored together with CRH in hypothalamic cells (Grossman, 1988).

When considering the possible role of opioids in stress or stereotypic behaviour, it is necessary to distinguish between a central and a peripheral pool of opioids. Although opioids can cross the blood–brain barrier (Banks and Kastin, 1990), it is not clear whether peripheral opioids play a role in the CNS. Similarly, opioids can be released locally in specific brain areas without being detectable in peripheral blood. Accordingly, different CSF and plasma concentrations can be found during stress (Smith et al., 1986; Owens et al., 1988)

For obvious reasons, most investigations on stress are based on the collection of blood samples from peripheral vessels and the determination of the plasma concentration of some hormone (e.g. glucocorticosteroid, catecholamine) under the assumption that changes in the concentration of these hormones reflect processes of the CNS. From the foregoing, however, it becomes obvious that, for instance, glucocorticosteroid release is no direct measure of CRH release, but that the release is modulated by a number of intervening factors, such as vasopressin, catecholamines, endogenous opioids, etc. Stated differently, the more peripheral to the brain the stress indicators are measured, the greater is the involvement of modulating factors.

In addition to the already mentioned sources of variability, it is necessary to take various stress-independent factors into consideration before changes in a hormone release can be used as a stress indication. These stress-independent factors concern primarily the basal secretory pattern of the hormones, their transport in blood, and their disappearance rate.

Like most other hormones, CRH, ACTH and glucocorticosteroids are secreted in an episodic (or pulsatile) fashion (Hellman et al., 1970; Gallagher et al., 1973; Ixart et al., 1987). In addition to this ultradian variability, secretory episodes are unevenly distributed over the 24-hour period giving rise to the circadian variation. In diurnal animals, most secretory episodes occur between the early morning (the latter part of the sleep phase) and the early afternoon. In nocturnal animals the circadian rhythm is reversed. The photoperiod provides the most important entrainment for the circadian rhythm (Czeisler et al., 1976). Because the signal of a hormone (such as the 'stress signal') is transmitted by relative changes in its secretory pattern (e.g. relative increase or decrease in its blood concentration or the interval between such changes) rather than its absolute plasma level (Yates, 1981), it is necessary to collect blood sufficiently frequently (for cortisol every 10 to 20 minutes) and over a sufficiently long period (e.g. 24 hours), to determine the hormone concentration in each sample, and to construct a secretory profile. The episodic secretory pattern can then be analysed objectively for episode frequency, amplitude and interval by special computer programs (e.g. Merriam and Wachter, 1982; Veldhuis and Johnson, 1986).

In acute stress situations, in which the stress reactions are intense enough to clearly surpass this circadian and ultradian variation, determination of the episodic secretion is less important. In chronic or chronic intermittent stress situations, however, in which some degree of adaptation may have occurred (e.g. Becker et al., 1985, 1989; Ladewig and Smidt, 1989), a stress-induced change in the secretion of a hormone may only be discernible after a detailed analysis of the secretory pattern and comparison with that of unstressed animals (Ladewig and Smidt, 1989).

For other hormones, additional intervening factors may be in operation. For the catecholamines, for instance, different production sites are responsible for different secretory rates both during basal secretion and during stress. Adrenaline is primarily released as a hormone from the adrenal medulla, whereas noradrenaline is primarily released as a neurotransmitter in the peripheral nervous system. Its presence in the blood stems from a 'spillover' from the synapses (Halter et al., 1980). Consequently, stress reactions that include muscular activity affect noradrenaline relatively more than adrenaline, whereas psychological stress (e.g. mental stress in humans) affects adrenaline relatively more than noradrenaline.

During blood transport most hormones are, to some extent, reversibly bound to carrier proteins. These proteins are either hormone-specific proteins with high affinity (e.g. corticosteroid binding globulin or transcortin) or non-specific with low affinity, that is they are able to bind several different hormones (e.g. albumin). The binding of a hormone protects it from metabolism, thereby acting as a storage and buffer system for the hormone. Since only the unbound free portion of a hormone is biologically active, however, it is necessary to take the binding rate into consideration before the 'stress signal' of a hormone can be evaluated (Barnett et al., 1981; Follenius and Brandenberger, 1986).

Finally, the plasma concentration of a hormone depends not only on the production of the hormone but also on its disappearance (e.g. uptake in target organs, metabolism in the liver, and excretion with urine, milk, sweat, saliva, etc.) and these processes may vary throughout the 24-hour period (e.g. De Lacerda et al., 1973). It must be recognized that the metabolic clearance rate may also be affected by stress. Again, consideration of these factors is probably less essential in acute type stress reactions but could affect results of chronic and chronic intermittent type studies.

In the following sections, we consider the relationships between the performance of stereotypic behaviour and physiological systems involved in stress.

Stereotypies and Endogenous Opioids

An inhibitory effect of opioid antagonists on stereotypic behaviour has been found in pigs (Cronin *et al.*, 1985; Rushen *et al.*, 1990), poultry (Savory *et al.*, 1992), voles (Kennes *et al.*, 1988), and horses (Dodman *et al.*, 1987). In the study of Cronin *et al.* (1985), naloxone was found to reduce stereotypies by more than 50%. Cronin *et al.* also found the effects of naloxone to be specific to stereotypic behaviour: naloxone did not affect other behaviours such as exploratory behaviour in loose-housed sows. This last result is somewhat at variance with work on rodents showing that endogenous opioids are implicated in a wide range of behaviours (Rodgers and Cooper, 1988). In more recent work, Rushen *et al.* (1990) and Schouten and Rushen (1992) found that after injecting tethered, stereotyping sows with naloxone, non-stereotypic behaviours, particularly drinking, were also affected by naloxone. In addition, sows kept in individual pens with deep straw, and showing little or no stereotypic behaviour, also decreased non-stereotypic manipulation of straw after treatment with naloxone, confirming that the effects of naloxone are not necessarily specific to stereotypic behaviour (Schouten and Rushen, 1992). Furthermore naloxone reduced the time sows spent performing stereotypic behaviour by only 20–30% (Rushen *et al.*, 1990), or by 10% (Schouten and Rushen, 1992), considerably less than that reported by Cronin (1985). Last, treatment with naloxone actually increased the time spent barbiting, which was seen infrequently in the control condition, suggesting that some apparently stereotypic activities may be increased by naloxone (Schouten and Rushen, 1992).

One difference between these studies may have been the respective stage of development of the stereotypy. Cronin (1985), using sows that had performed stereotypies between 3 and 53 weeks, found a negative correlation between time since development of the stereotypy and the inhibiting effect of naloxone. A similar negative correlation has subsequently been found in a study on stereotypically jumping voles, where the effect of naloxone was also decreased by increasing age of stereotypy (Kennes *et al.*, 1988). In contrast, Schouten and Rushen (1992) found that naloxone-induced inhibition of stereotypies was similar at both 1 and 2 months, nor was the effect stronger in sows showing more intense stereotypic behaviour, indicating that the effects of naloxone are not necessarily greater for less developed stereotypies.

An alternative to using opioid antagonists is to look for direct evidence of endogenous opioid activity when pigs are performing stereotypic behaviour. Rushen *et al.* (1990) found evidence of an opioid-based reduction in sensitivity to pain after the feeding period. Although stereotypies occurred mainly during the post-feeding period, sows performing little stereotypic behaviour showed the largest reduction in sensitivity to pain, rather suggesting a negative correlation between stereotypies and opioid activity. While in this study it was not clear whether any causal relationships were involved, recently,

Zanella *et al.* (1991) found that sows showing stereotypic behaviour had a lower density of κ opioid receptors in the frontal cortex than sows showing little stereotypic behaviour, while tethered sows had a higher density of μ opioid receptors than group-housed sows. These interesting results need to be extended.

Involvement of endogenous opioids in stereotypic behaviour could be related to the sustenance of stereotypic behaviour by enhancing dopaminergic activity underlying the stereotypic behaviour (Dantzer, 1986; see also Chapter 6). Alternatively since opioids have been implicated in reward processes, the persistence of stereotypic behaviour could be due to opioid-based reinforcement (Dantzer, 1986; Rushen *et al.*, 1990; see Dantzer, 1991, and Mason, 1991, for a discussion of this hypothesis). Rushen *et al.* (1990) found that naloxone reduced the duration of bouts of behaviour, while increasing the frequency of bouts, thus increasing switching between different behaviours. Consequently, the authors have suggested that naloxone reduced total time spent in stereotypic behaviour mainly by reducing the mean duration of single bouts of stereotypic behaviour which could reflect on a reduction on opioid-based reinforcement of the behaviour.

In conclusion, opioid antagonists reliably reduce time spent in stereotypic behaviour, suggesting that endogenous opioids are involved in performance of stereotypies. However, the effects of naloxone are not specific to stereotypic behaviour or to animals showing high levels of stereotypies, and may be more generally related to reduction of the persistence of behaviour. The complexity of the relationship between stereotypic behaviour and endogenous opioids needs to be acknowledged.

Stereotypies and the Hypothalamic–Pituitary–Adrenal Axis

The conditions under which stereotypies appear are often seen as involving high levels of arousal or stress; it has therefore been suggested that the performance of stereotypies may be a way of reducing the impact of these conditions on the animal's psychological or physiological state (see Chapter 7). A change in the activity of the hypothalamic–pituitary–adrenal (HPA) axis during the performance of stereotypic behaviour would therefore be a logical consequence.

Research on laboratory rodents has not produced a clear demonstration that stereotypies reduce HPA activity (Chapter 7). Lower concentrations of corticosteroids were found in poultry showing stereotypic spot-pecking behaviour (Kostal *et al.*, 1992), and similarly in pigs' levels of stereotypies were found to be negatively correlated to free corticosteroid concentrations (Cronin and Barnett, 1987), although both these studies took only small numbers of blood samples per individual. Evidence from other studies on pigs is less clearcut. In a study on individually confined sows, von Borell and Hurnik (1991) found no differences in levels of circulating cortisol between stereotyping and non-stereotyping individuals. In the same study, stereotyping individuals exhibited a higher adrenocortical response to ACTH challenge than sows that did not perform stereotypies (von Borell and Hurnik, 1991). Since (non-stereotyping) pigs show a high variability in adrenocortical activity (Dantzer and Mormède, 1983a; Ladewig, 1987; von Borell and Ladewig, 1989), little can be deduced about the direction of the causal relationships, if any, at this stage. In a recent study Terlouw et al. (1991) collected blood samples from tethered sows which showed large individual differences in the daily amount of chain-manipulation and drinking. Blood samples were regularly taken over several days around feeding time, with or without the chain present, and average plasma cortisol concentrations were compared with time spent manipulating the chain and drinking in a correlational analysis. Both chain-manipulation and drinking occurred at high levels after feeding. Although plasma cortisol levels were also high around feeding, sows that spent a large amount of time manipulating the chain did not differ in plasma cortisol levels from their non-chain-manipulating counterparts. Furthermore, short-term and long-term chain removal did not increase plasma cortisol levels of chain-manipulating sows. However, cortisol levels were positively correlated across days that chains were removed, but not across control days. This result indicates that sows were consistent in their response to the prevention of chain-manipulation. It is not clear, however, whether the cortisol reaction was a response to prevention of chain-manipulation or to the general reduction in noise levels that accompanied chain removal. A detailed analysis of the data further revealed that, within individuals, manipulation of the chain was correlated to lower levels of plasma cortisol on a temporal basis. However, the effect of chain-manipulation on plasma cortisol concentrations was inconsistent and accounted only for 5% of the variation, suggesting that the effect was very small (Terlouw et al., unpublished results). In contrast, a consistent negative correlation between the daily amount of drinking and average daily plasma cortisol concentrations was found. These data need not necessarily indicate that drinking reduced the sows' stress levels, as the physiological consequences of the ingestion of large amounts of water (e.g. reduced synergistic action of vasopressin on CRH-induced ACTH release) may be responsible for the drinking-induced reduction in plasma cortisol, rather than a reduction in stress intensity.

Stereotypic and adjunctive (or schedule-induced) activities have been compared to each other on a behavioural level, as both include a routine type activity and both develop over time under conditions of apparent stress (Rushen, 1984; Dantzer, 1986; Dantzer et al., 1988; Terlouw et al., 1991). In a series of experiments, Dantzer and co-workers investigated the effect of schedule-induced nibbling and chewing on a composite 'toy' (a metal chain covered with cloth strips, rubber hose, leather and nylon straps), on plasma cortisol levels in post-weaning pigs (Dantzer and Morméde, 1981, 1983b; Dantzer et al., 1987). Pigs that were trained to 'chain-chew' during the intervals of an intermittent food-delivery schedule showed reduced plasma cortisol levels after the session, as compared to before. Removal of the chain prevented this reduction. However, the apparent reduction in plasma cortisol in response to chain-chewing was due to an increase in cortisol levels before the session; pigs that had been trained on sessions of intermittent food delivery without a chain had similar cortisol levels after the session as chain-chewing pigs. In contrast to these results, no relationship between level of chain-manipulation and pre-feeding levels of cortisol was found in tethered sows by Terlouw et al. (1991). The finding that chain-manipulation following delivery of food in a massed form rather than intermittently was accompanied by a rise in plasma cortisol, further suggests that stereotypic and adjunctive chain-manipulation are not entirely equivalent (Dantzer et al., 1987).

Thus generally data on a relationship between performance of stereotypies and HPA-activity are inconsistent, with examples of both positive and negative effects. While the causes of these differences may be manyfold, one obvious possibility is the lack of consistency in comparisons between developing, and established stereotypies.

Stereotypies and the Autonomic Nervous System

The main function of the autonomic nervous system (ANS) is to maintain homeostasis. The ANS regulates the activities of many organs and helps visceral functions adapt to the animal's immediate needs. The ANS consists of two major divisions, the sympathetic and parasympathetic division. Many organs are innervated by both divisions which usually exert antagonistic effects. For example, activity in the sympathetic innervation to the heart leads to increased heart rate, while activity in the parasympathetic innervation decreases heart rate. Thus, increases in heart rate can result from either increased sympathetic or decreased parasympathetic activity.

Sympathetic and parasympathetic activity is coordinated in the brain stem. Information is received primarily from the limbic system and the neocortex (Fig. 5.2), both of which play a crucial role in the perception and evaluation of stressors and the organization of stress responses. In the upper brain stem, particularly in the nucleus solitarius, this information is transformed into signals that are transmitted via numerous nuclei of the lower brain stem, regulating the activity of preganglionic neurons of the sympathetic and parasympathetic systems. Endogenous opioids play an important role in modulating autonomic activity. Recent studies have shown that opioid reduction of heart rate involves μ opioid receptor-mediated increases in parasympathetic activity (Holaday, 1983; Grossman, 1988; Marson *et al.*, 1989; Morris *et al.*, 1990).

It is often suggested that stereotypies occur in response to a high level of arousal. A link between stereotypies and arousal in pigs was suggested by the finding that stereotypic behaviour and sympathetic nervous activity increase in the post-feeding period (Schouten *et al.*, 1991). In other species, feeding has been found to induce an increased heart rate (Bloom *et al.*, 1975; Matsukawa and Ninomija, 1987; Muller *et al.*, 1989). However, Schouten *et al.* (1991) also identified an effect of environment as heart rate at feeding was found to be higher in tethered sows compared to loose-housed sows; this effect was blocked by the beta-adrenoreceptor blocker, carazolol, indicating sympathetic nervous involvement. The difference between tethered and loose-housed animals increased with the duration of tethering. Stereotypic behaviour also occurred more in tethered sows, particularly during the post-feeding period (e.g. Rushen, 1985). However, reducing heart rate by beta-blockers did not reduce stereotypic behaviour, suggesting that the behaviours are not responses to the increased heart rate itself. The opioid antagonist naloxone reduced stereotypic behaviour and led to consistent increases in heart rate (Schouten *et al.*, 1991).

The relationships between individual levels of stereotypic behaviour and heart rate are not consistent over time in sows. For instance, sows tethered for 1 month showed no correlation between the level of stereotypies and that of heart rate. After 2 months of tethering, sows showing stereotypic behaviour had lower heart rate at feeding than others. After 6–8 months of tethering, these sows had slightly higher heart rates than others, although this result may have reflected their higher activity levels (Schouten and Wiepkema, 1991; Schouten *et al.*, 1991).

In children, stereotypic leg-swinging has been found to be associated with reduced heart rate (Soussignan and Koch, 1985), while in pigs, heart rates tended also to be lower during bouts of stereotypic behaviour (Schouten and Wiepkema, 1991). Although this observation might suggest that the performance of stereotypies reduces heart rate. Preventing stereotypic behaviour in long-term tethered sows by removing the chains led to a slight decrease in heart rate rather than an increase. The performance of stereotypies was not necessary for the reduction of heart rate that occurred after feeding (Schouten *et al.*, 1991).

Conditions that lead to stereotypic behaviour in pigs also lead to increased sympathetic nervous activity, especially at feeding. As yet, there is insufficient evidence, however, to infer the causal effects underlying this relationship. The association between stereotypic behaviour and the autonomic nervous system clearly requires more study.

Stereotypies and other Physiological Systems

Since stress is known to have marked effects on reproduction (Moberg, 1987; Knol, 1991; Rivier and Rivest, 1991) it would be logical to look for associations between stereotypic behaviour and reproductive function. Redbo *et al.* (1992) found that high-yielding dairy cows showed higher levels of stereotypic tongue-play. Most other studies on stereotypic behaviour and reproduction have been done on pigs but results have not been consistent. Confinement of sows and gilts at mating can reduce conception rate and perhaps litter size (Hemsworth and Barnett, 1989; Barnett and Hemsworth, 1991). Such confinement favours the development of stereotypic behaviour. Cronin (1985) reported that, for lower parity sows, high levels of stereotypic behaviour were associated with larger litters while the relationship was reversed among older sows. McGlone and Blecha (1987) found that young sows showing stereotypic behaviour had larger litters. Von Borell and Hurnik (1990) found that, although sows performing stereotypic behaviour gave birth to fewer live piglets than sows not performing such behaviour, within the group of stereotyping sows there was a positive correlation between the amount of stereotypic behaviour and litter size. All of these studies, however, are correlational so that the direction of the causal relationships cannot be determined. Furthermore, relationships between stereotypic behaviour and the endocrine systems underlying reproduction have not been extensively investigated.

Stress can also affect physiological systems involved in digestion (e.g. Williams *et al.*, 1987; Monnikes *et al.*, 1992). Wiepkema *et al.* (1987) reported that calves showing stereotypic behaviour had a lower incidence of abomasal ulceration, although this finding has not been confirmed (Lidfors, 1992). In view of the close association between stereotypic behaviour and feeding (see Chapter 3), an association between such behaviour and the physiological processes involved in digestion could be expected. The idea that the expression or inhibition of a behaviour could influence digestive hormones has received some support from studies on non-nutritive sucking in infants and calves. Sucking a pacifier has been found to have a quieting effect on infants (Kessen and Leutzendorff, 1963; Field and Goldson, 1984) and a growth-promoting effect via increased insulin and gastrin secretion and decreased somatostatin secretion as well as altered gastric motor function that follows the non-nutritive sucking (Marchini *et al.*, 1987; Widstrom *et al.*, 1988). In calves, sucking a non-functional rubber teat after feeding resulted in increased secretion of insulin and possibly a changed pattern of secretion of cholecystokinin (de Passillé *et al.*, 1991). Clearly, more research is needed on the possible relationship between stereotypic behaviour and digestive function.

Conclusion

From the above evidence, a clear picture of the relationship between stereotypic behaviour and neuroendocrine systems involved in stress reactions cannot yet be drawn. The heterogeneity of stereotypies, the way each animal perceives its environment, and the complex control of the neuroendocrine systems are probably the main reasons for the inconsistency of the results. While some studies suggest that the performance of stereotypies may serve to reduce HPA and ANS activity, other studies contradict this. Clearly, the relationship between stereotypies and the neuroendocrine system is too complex to fit the simple form of the 'coping' hypothesis. The role of CRH in stereotypic behaviour has, so far, not been extensively investigated. A shift from measurements of peripheral physiological activity to measurements of neuroendocrine activity in higher brain regions might be more promising for understanding the relationship between stress and stereotypic behaviours.

References

Akil, H., Watson, S.J., Young, E., Lewis, M.E., Khachaturian, H. and Walker, J.M. (1984) Endogenous opioids: biology and function. *Annual Review of Neuroscience* 7, 223–255.

Amir, S., Brown, Z.W. and Amit, Z. (1980) The role of endorphins in stress. Evidence and speculations. *Neuroscience Biobehavioural Review* 4, 77–86.

Amit, Z. and Galina, Z.H. (1986) Stress-induced analgesia: adaptive pain suppression. *Physiological Reviews* 66, 1091–1120.

Axelrod, J. and Reisine, T.D. (1984) Stress hormones: their interaction and regulation. *Science* 224, 458–459.

Banks, W.A. and Kastin, A.J. (1990) Peptide transport systems for opiates across the blood-brain barrier. *American Journal of Physiology* 259, E1–E10.

Barnett, J.L. and Hemsworth, P.H. (1991) The effects of individual and group housing on sexual behaviour and pregnancy in pigs. *Animal Reproductive Science* 25, 265–273.

Barnett, J.L., Winfield, C.G., Cronin, G.M. and Makin, A.W. (1981) Effects of photoperiod and feeding on plasma corticosteroid concentrations and maximum corticosteroid-binding capacity in pigs. *Australian Journal of Biological Science* 34, 577–585.

Becker, B.A., Nienaber, J.A., DeShazer, J.A. and Hahn, G.L. (1985) Effect of transportation on cortisol concentrations and on the circadian rhythm of cortisol in gilts. *American Journal of Veterinary Research* 46, 1457–1459.

Becker, B.A., Christenson, R.K., Ford, J.J., Nienaber, J.A., DeShazer, J.A. and Hahn, G.L. (1989) Adrenal and behavioural responses of swine restricted to varying degrees of mobility. *Physiology and Behaviour* 45, 1171–1176.

Benus, R.F. (1988) Aggression and coping. PhD Thesis, University of Gronigen, The Netherlands.

Bloom, S.R., Edwards, A.V., Hardy, R.M., Malinowska, K. and Silver, M. (1975) Cardiovascular and endocrine responses to feeding in young calves. *Journal of Physiology* 253, 135–155.

Burchfield, S.R. (1979) The stress response: a new perspective. *Psychosomatic Medicine* 41, 661–672.

Cannon, W.B. (1929) *Bodily Changes in Pain, Hunger, Fear and Rage: An Account of Recent Researches into the Function of Emotional Excitement*, 2nd edition. Appleton, New York.

Cronin, G.M. (1985) The development and significance of abnormal stereotyped behaviours in tethered sows. PhD Thesis, University of Wageningen, The Netherlands.

Cronin, G.M. and Barnett, J.L. (1987) An association between plasma corticosteroids and performance of stereotypic behaviour in tethered sows. In: Barnett J.L. *et al.* (eds), *Manipulating Pig Production*. Australian Pig Science Association, Werribee, Australia, p. 26.

Cronin, G.M., Wiepkema, P.R. and van Ree, J.M. (1985) Endogenous opioids are involved in abnormal stereotype behaviours of tethered sows. *Neuropeptides* 6, 527–530.

Czeisler, C.A., Ede, M.C.M., Regestein, Q.R., Kisch, E.S., Fang, V.S. and Ehrlich, E.N. (1976) Episodic 24-hour cortisol secretory patterns in patients awaiting elective cardiac surgery. *Journal of Clinical Endocrinology and Metabolism* 42, 273–283.

Dantzer, R. (1986) Behavioural, physiological and functional aspects of stereotyped behaviour: a review and a reinterpretation. *Journal of Animal Science* 62, 1776–1786.

Dantzer, R. (1991) Stress, stereotypies and welfare. *Behavioural Processes* 25, 95–102.

Dantzer, R. and Mormède, P. (1981) Pituitary-adrenal consequences of adjunctive activities in pigs. *Hormones and Behaviour* 15, 386–395.

Dantzer, R. and Mormède, P. (1983a) Stress in farm animals: A need for reevaluation. *Journal of Animal Science* 57, 6–18.

Dantzer, R. and Mormède, P. (1983b) De-arousal properties of stereotyped behaviour: evidence from pituitary adrenal correlates in pigs. *Applied Animal Ethology* 10, 233–244.

Dantzer, R., Gonyou, H.W., Curtis, S.E. and Kelley, K.W. (1987) Changes in serum cortisol reveal functional differences in frustration-induced chain-chewing in pigs. *Physiology and Behaviour* 39, 775–777.

Dantzer, R., Terlouw, C., Mormède, P. and Le Moal, M. (1988) Schedule-induced polydipsia experience decreases plasma corticosterone levels but increases plasma prolactin levels. *Physiology and Behaviour* 43, 275–279.

De Boer, S.F., Slangen, J.L. and van der Gugten, J. (1988) Adaptation of plasma catecholamine and corticosterone responses to short-term repeated noise stress in rats. *Physiology and Behaviour* 44, 273–280.

De Boer, S.F., Koopmans, S.J., Slangen, J.L. and van der Gugten, J. (1990) Plasma catecholamine, corticosterone and glucose responses to repeated stress in rats: effect of interstressor interval length. *Physiology and Behaviour* 47, 1117–1124.

De Lacerda, L., Kowarski, A. and Migeon, C.J. (1973) Integrated concentration and diurnal variation of plasma cortisol. *Journal of Clinical Endocrinology and Metabolism* 36, 227–238.

de Passillé, A.M.B., Christopherson, R.J. and Rushen, J. (1991) Sucking behaviour affects post-prandial secretion of digestive hormones in the calf. In: Appleby, M.C., Horrell, R.I., Petherick, J.C. and Rutter, S.M. (eds), *Applied Animal Behaviour: Past, Present and Future*. UFAW, Potters Bar, pp. 130–131.

Dodman, N.H., Shuster, L., Court, M.H. and Dixon, R. (1987) Investigation into the use of narcotic antagonists, in the treatment of a stereotypic behaviour pattern (crib-biting) in the horse. *American Journal of Veterinary Research* 48, 311–319.

Dunn, A.J. and Berridge, C.W. (1990) Physiological and behavioural responses to corticotropin-releasing factor administration: is CRF a mediator of anxiety or stress responses? *Brain Research Review* 15, 71–100.

Field, T. and Goldson, E. (1984) Pacifying effects of nonnutritive sucking on term and preterm neonates during heelstick procedures. *Pediatrics* 74, 1012–1015.

Follenius, M. and Brandenberger, C.W. (1986) Plasma free cortisol during secretory episodes. *Journal of Clinical Endocrinology and Metabolism* 62, 609–612.

Gallagher, T.F., Yoshida, K., Roffwarg, H.D., Fukushima, D.K., Weitzman, E.D. and Hellman, L. (1973) ACTH and cortisol secretory patterns in man. *Journal of Clinical Endocrinology and Metabolism* 36, 1058–1068.

Grossman, A. (1988) Opioids and stress in man. *Journal of Endocrinology* 119, 377–381.

Halter, J.B., Pflug, A.E. and Tolas, A.G. (1980) Arterial-venous differences of plasma catecholamines in man. *Metabolism* 29, 9–12.

Hellman, L., Nakada, F., Curti, J., Weitzman, E.D., Kream, J., Roffwarg, H., Ellman, S., Fukushima, D.K. and Gallagher, T.F. (1970) Cortisol is secreted episodically by normal man. *Journal of Clinical Endocrinology* 30, 411–422.

Hemsworth, P.H. and Barnett, J.L. (1989) Behavioural responses affecting gilt and sow reproduction. *Journal of Reproduction and Fertility (Suppl.)* 40, 343–354.

Hennessy, D.P., Stelmasiak, T., Johnston, N.E., Jackson, P.N. and Outch, K.H. (1988) Consistent capacity for adrenocortical response to ACTH administration in pigs. *American Journal of Veterinary Research* 49, 1276–1283.

Henry, J.P. and Stephens, P.M. (1977) *Stress, Health, and the Social Environment. A Sociobiologic Approach to Medicine. Topics in Environmental Physiology and Medicine*. Springer-Verlag, New York.

Holaday, J.W. (1983) Cardiovascular effects of endogenous opiate systems. *Annual Review of Pharmacology and Toxicology* 23, 54–91.

Ixart, G., Barbanel, G., Conte-Devolx, B., Grin, M., Oliver, C. and Assenmacher, I. (1987) Evidence for basal and stress-induced release of corticotropin releasing factor in the push-pull cannulated median eminence of conscious free-moving rats. *Neuroscience Letters* 74, 85–89.

Kennes, D., Ödberg, F.O., Bouquet, Y. and de Rycke, P.H. (1988) Changes in naloxone and haloperidol effects during the development of captivity-induced jumping stereotypy in bank voles. *European Journal of Pharmacology* 153, 19–24.

Kessen, W. and Leutzendorff, A.-M. (1963) The effect of non-nutritive sucking on movement in the human newborn. *Journal of Comparative and Physiological Psychology* 56, 69–72.

Knol, B.W. (1991) Stress and the endocrine hypothalamus-pituitary-testis system: a review. *Veterinary Quarterly* 13, 104–113.

Konarska, M., Stewart, R.E. and McCarty, R. (1990) Predictability of chronic intermittent stress: effects on sympathetic-adrenal medullary responses of laboratory rats. *Behavioural and Neural Biology* 53, 231–243.

Kostal, L., Savory, C.J. and Hughes, B.O. (1992) Diurnal and individual variation in behaviour of restricted-fed broiler breeders. *Applied Animal Behaviour Science* 32, 361–374.

Ladewig, J. (1987) Endocrine aspects of stress: Evaluation of stress reactions in farm animals. In: Wiepkema, P.R. and van Adrichem, P.W.M. (eds), *Biology of Stress in Farm Animals: An Integrative Approach*. Martinus Nijhoff, Dordrecht, pp. 13–25.

Ladewig, J. and Smidt, D. (1989) Behavior, episodic secretion of cortisol, and adrenocortical reactivity in bulls subjected to tethering. *Hormones and Behavior* 23, 344–360.

Ladewig, J. and von Borell, E. (1988) Ethological methods alone are not sufficient to measure the impact of environment on animal health and animal well-being. In: Unshelm, J., van Putten, G., Zeeb, K. and Ekesbo, I. (eds) *Proceedings of the International Congress on Applied Ethology in Farm Animals, Skara 1988*. Kuratorium für Technik und Bauwesen in der Landwirtschaft, Darmstadt, pp. 95–102.

Lawrence, A.B., Terlouw, E.M.C. and Illius, A.W. (1991) Individual differences in behavioural responses of pigs exposed to non-social and social challenges. *Applied Animal Behaviour Science* 30, 73–86.

Lewis, J., Mansour, A., Khachaturian, H., Watson, S.J. and Akil, H. (1987) Opioids and pain regulation. In: Akil, H. and Lewis, J.W. (eds), *Neurotransmitters and Pain Control*. Karger, Basel, pp. 129–159.

Lidfors, L. (1992) Behaviour of bull calves in two different housing systems. Thesis. Swedish University of Agricultural Sciences, Skara.

Lyons, D.M., Price, E.O. and Moberg, G.P. (1988) Social modulation of pituitary-adrenal responsiveness and individual differences in behaviour of young domestic goats. *Physiology and Behaviour* 43, 451–458.

Marchini, G., Lagercrantz, H., Feuerberg, Y., Winberg, J. and Üvnas-Moberg, K. (1987) The effect of non-nutritive sucking on plasma insulin, gastrin and somatostatin in newborns. *Acta Pediatrica Scandinavia* 76, 573–576.

Marson, L., Kiritsy-Poy, J.A. and van Loon, G.R. (1989) Opioid peptide modulation of cardiovascular and sympathoadrenal responses to stress. *American Journal of Physiology* 257, R901–R908.

Mason, G.J. (1991) Stereotypies: a critical review. *Animal Behaviour* 41, 1015–1037.

Mason, J.W. (1974) Specificity in the organization of neuroendocrine response profiles. In: Seeman, P. and Brown, G. (eds), *Frontiers in Neurology and Neuroscience Research*. University of Toronto, pp. 68–80.

Matsukawa, K. and Ninomija, I. (1987) Changes in renal sympathetic nerve activity, heart rate and arterial blood pressure associated with eating in cats. *Journal of Physiology* 390, 229–242.

McGlone, J.J. and Blecha, F. (1987) An examination of behavioural, immunological and productive traits in four management systems for sows and piglets. *Applied Animal Behaviour Science* 18, 269–286.

Merriam, G.R. and Wachter, K.W. (1982) Algorithms for the study of episodic hormone secretion. *American Journal of Physiology* 243, E310–E318.

Moberg, G. (1987) Influence of the adrenal axis upon the gonads. *Oxford Review of Reproductive Biology* 9, 456–496.

Monnikes, H., Schmidt, B.G., Raybould, H.E. and Taché, Y. (1992) CRF in the paraventricular nucleus mediates gastric and colonic motor responses to restraint stress. *American Journal of Physiology* 262, G137–G143.

Morris, M., Salmon, P., Steinberg, H., Sykers, E.A., Bouloux, P., Newbould, E., McLoughlin, L., Besser, G.M. and Grossman, A. (1990) Endogenous opioids modulate the cardiovascular response to mental stress. *Psychoneuroendocrinology* 15, 185–192.

Muller, C., Ladewig, J., Thielscher, H.-H. and Smidt, D. (1989) Behavior and heart rate of heifers housed in tether stanchions without straw. *Physiology and Behavior* 46, 751–754.

Natelson, B.H., Ottenweller, J.E., Cook, J.A., Pitman, D., McCarty, R. and Tapp, W.N. (1988) Effects of stressor intensity on habituation of the adrenocortical stress response. *Physiology and Behavior* 43, 41–46.

Owens, P.C., Chan, E.C., Lovelock, M., Falconer, J. and Smith, R. (1988) Immunoreactive methionine-enkephalin in cerebrospinal fluid and blood plasma during acute stress in conscious sheep. *Endocrinology* 122, 311–318.

Pitman, D.L., Ottenweller, J.E. and Natelson, B.H. (1990) Effect of corticoid responses in rats. *Behavioural Neuroscience* 104, 28–36.

Redbo, I., Jacobsson, K.G., van Doorn, C. and Pettersson, G. (1992) A note on the relations between oral stereotypies in dairy cows and milk production, health and age. *Animal Production* 54, 166–168.

Rivier, C. and Rivest, S. (1991) Effect of stress on the activity of the hypothalamic-pituitary-gonadal axis: peripheral and central mechanisms. *Biology of Reproduction* 45, 523–532.

Rodgers, R.J. and Cooper, S.J. (1988) *Endorphins, Opiates and Behavioural Processes*. Wiley, New York.

Rushen, J. (1984) Stereotyped behaviour, adjunctive drinking and the feeding periods of tethered sows. *Animal Behaviour* 32, 1059–1067.

Rushen, J. (1985) Stereotypies, aggression and the feeding schedules of tethered sows. *Applied Animal Behavioural Science* 14, 137–147.

Rushen, J. (1991) Problems associated with the interpretation of physiological data in the assessment of animal welfare. *Applied Animal Behavioural Science* 28, 381–386.

Rushen. J., de Passillé, A.M. and Schouten, W. (1990) Stereotyped behaviour, endogenous opioids and post-feeding hypoalgesia in pigs. *Physiology and Behaviour* 48, 91–96.

Savory, C.J., Seawright, E. and Watson, A. (1992) Stereotyped behaviour in broiler breeders in relation to husbandry and opioid receptor blockade. *Applied Animal Behavioural Science* 32, 349–360.

Schouten, W. and Rushen, J. (1992) Effects of naloxone on stereotypic and normal behaviour of tethered and loose houses sows. *Applied Animal Behavioural Science* 33, 17–26.

Schouten, W.G.P. and Wiepkema, P.R. (1991) Coping styles of tethered sows. *Behavioural Processes* 25, 125–132.

Schouten, W., Rushen, J. and de Passillé, A.M. (1991) Stereotypic behaviour and heart rate in pigs. *Physiology and Behaviour* 50, 617–624.

Selye, H. (1956) *The Stress of Life*. McGraw-Hill, New York.

Smidt, D. (1983) Advantages and problems of using integrated systems of indicators as compared to single traits. In: Smidt, D. (ed), *Indicators Relevant to Farm Animal*

Welfare. *Current Topics in Veterinary Medicine and Animal Science.* Martinus Nijhoff, Dordrecht, pp. 201–207.

Smith, R., Owens, P.C., Lovelock, M., Chan, E.-C. and Falconer, J. (1986) Acute hemorrhagic stress in conscious sheep elevates immunoreactive β-endorphin in plasma but not in cerebrospinal fluid. *Endocrinology* 118, 2572–2576.

Soussignan, R. and Koch, P. (1985) Rhythmical stereotypies (leg-swinging) associated with reduction in heart-rate in normal school children. *Biological Psychology* 21, 161–167.

Székely, J.I. and Rónai, A.Z. (eds) (1982) *Opioid Peptides. Volume II. Pharmacology.* CRC Press, Boca Raton, Florida.

Terlouw, E.M.C., Lawrence, A.B., Ladewig, J., de Passillé, A.M.B., Rushen, J. and Schouten, W. (1991) A relationship between stereotypies and cortisol in sows. *Behavioural Processes* 25, 133–153.

Veith-Flanigan, J. and Sandman, C.A. (1985) Neuroendocrine relationships with stress. In: Burchfield, S.R. (ed.), *Stress, Psychological and Physiological Interactions.* Hemisphere, Washington DC, pp. 129–161.

Veldhuis, J.D. and Johnson, M.L. (1986) Cluster analysis: a simple, versatile and robust algorithm for endocrine pulse detection. *American Journal of Physiology* 250, E486–E493.

von Borell, E. and Hurnik, J.F. (1990) Stereotypic behavior and productivity of sows. *Canadian Journal of Animal Science* 70, 953–956.

von Borell, E. and Hurnik, J.F. (1991) Stereotypic behavior, adrenocortical function, and open field behavior of individually-confined gestating sows. *Physiology and Behavior* 49, 709–714.

von Borell, E. and Ladewig, J. (1989) Altered adrenocortical response to acute stressors or ACTH(1–24) in intensely housed pigs. *Domestic Animal Endocrinology* 64, 299–309.

von Borell, E. and Ladewig, J. (1992) The relationship between behaviour and adrenocortical response pattern in domestic pigs. *Applied Animal Behavioural Science* 34, 195–206.

Widstrom, A.M., Marchini, G., Matthiesen, A.-S., Werner, S., Winberg, J. and Üvnas-Moberg, K. (1988) Nonnutritive sucking in tube fed preterm infants: effects on gastric motility and gastric contents of somatostatin. *Journal of Pediatric Gastroenterology and Nutrition* 7, 517–523.

Wiepkema, P.R., van Hellemond, K.K., Roessingh, P. and Romberg, H. (1987) Behaviour and abomasal damage in individual veal calves. *Applied Animal Behavioural Science* 18, 257–268.

Williams, C.L., Peterson, J.M., Villar, R.G. and Burks, T.F. (1987) Corticotropin-releasing factor directly mediates colonic responses to stress. *American Journal of Physiology (Gastrointestinal and Liver Physiology)* 253, G582–G586.

Yates, F.E. (1981) Analysis of endocrine signals: the engineering and physics of biochemical communication systems. *Biology of Reproduction* 24, 73–94.

Zanella, A.J., Broom, D.M. and Hunter, J. (1991) Changes in opioid receptors of sows in relation to housing, inactivity and stereotypies. In: Appleby, M.C., Horrell, R.I., Petherick, J.C. and Rutter, S.M. (eds), *Applied Animal Behaviour: Past, Present and Future.* UFAW, Potters Bar, pp. 140–141.

Neurobiological Basis of Stereotypies 6

SIMONA CABIB
Instituto di Psicobiologia e Psicofarmacologia (CNR),
Rome, Italy.

Editors' Introductory Notes:
It seems certain that our knowledge of stereotypies will be increased by understanding more of the brain mechanisms underlying the behaviour (see Chapter 5). There are, however, a number of limitations on applying the techniques of neuroscience to large animals under the housing conditions in which they develop stereotypies. For this reason it is important for those studying stereotypies in farm and zoo animals to be aware of advances in other areas and disciplines where such limitations are less important.

Cabib's experience of stereotypies is firmly grounded in neuroscience and the study of drug- (e.g. amphetamine) induced stereotypies in rats and mice. As she explains, the justification for this work is biomedical, being directed towards using such drug-induced behaviour as an animal model of schizophrenia. The result is an infinitely more detailed knowledge of the neurobiological basis of drug-induced stereotypies than has yet been achieved for the environment-induced stereotypies that are the main focus of this book. While we should be careful to acknowledge the differences between drug-induced and environment-induced stereotypies (e.g. Chapter 2), the data presented by Cabib clearly make a very important contribution to our attempts to understand the brain mechanisms underlying stereotypies in large animals. More than this, Cabib presents a synthesis in which she attempts to link the effects of stress with the incidence of environmentally induced stereotypy. Cabib uses the term 'stress' to represent that phase where the animal's attempts to adapt (or 'cope') with a change in its environment have failed

The central focus of Cabib's argument is the role of the brain's mesoaccumbens dopamine (DA) system, in both the neural origins of stereotypies and in adaptation to stress. DA agonists such as amphetamine appear to induce stereotypy by affecting the 'balance' between the nucleus accumbens (of the mesoaccumbens DA system) and the caudatus putamen (part of the nigrostriatal DA system), such that output from the caudatus occludes that from the nucleus accumbens. However, the anatomical position of the accumbens allows it to modulate activity in the caudatus (the reverse cannot occur), and it may continue to promote activity from the caudatus and thus be essential to performance of stereotypy. Cabib also notes that 'displacement' activities are dependent on an intact mesoaccumbens DA system.

The mesoaccumbens DA system also responds to a variety of 'stressors' including novelty and restraint. From an analysis of novelty induced climbing responses in mice, Cabib suggests that prolonged exposure to a 'stressor', such as novelty, 'normally' results in a stress-induced inhibition of mesoaccumbens DA activity, brought about by activation of the regulatory DA autoreceptors; this results in a reduction in the climbing response. Cabib proposes that severe or chronic stress (in combination with genetic factors) can result in a malfunction of the DA regulatory system, leading to a failure of the mesoaccumbens system to control its own activity. As a result of stress, the mesoaccumbens system may then become hyperresponsive to a wide variety of stimuli (the overt sign of this process being the phenomenon 'behavioural sensitization'). As indicated above, the mesoaccumbens system is involved in amphetamine-induced stereotypies; consequently in a chronically stressed animal the hyperresponsive mesoaccumbens DA system may be similarly critical in promoting stereotypic sequences of behaviour.

Cabib ends by pointing out that the role of the mesoaccumbens DA system in stress-induced sensitization and stereotypies is not exclusive, there being a large number of other brain structures (e.g. the hippocampus, the neocortex and the frontal cortex), and neuropeptides that can influence mesoaccumbens DA activity. For example, corticosterone (or cortisol) is necessary for the development of behavioural sensitization. Lastly, she suggests stereotypies may be more a 'side-effect' of the alterations to brain DA functioning that accompany stress, than an expression of coping.

Introduction

Neuroscientists are researchers with very different backgrounds who share a common interest in the study of the nervous system. For this reason neuroscientists may or may not be interested in behaviour; as in the case of those involved in mathematical modelling of neuronal circuits, in chemical interactions between neural systems or brain structures, or in cellular studies of the neuron. There are, however, cases of collaborative and coordinated efforts of these scientists towards the development of interdisciplinary approaches to specific behavioural phenomena. In many cases, the efforts spring from the need to understand the neurobiological basis of abnormal behaviour in humans.

This is not only due to the strong social interest and, consequently, to the economic support that is won by this kind of research, but also to a very fundamental need for neurobiological studies on human psychopathologies. Experimental research on neurobiology of abnormal behaviour in humans has to be conducted mainly on non-human subjects for methodological as well as ethical reasons. In order to extrapolate from the results obtained in this research to clinical studies in humans, experiments must be conducted within the framework of animal models that should bear both analogies and homologies with pathological behaviour in humans.

Stereotypic behaviour is a response that is readily induced by psychostimulants in the most commonly used laboratory species, the rat. Moreover, it has been suggested that it reproduces in animals a behavioural response that shares analogies with the stereotypic alteration of behaviour observed in several clinical studies of schizophrenic patients. Finally, a strong correlation has been shown between the therapeutic potency of the majority of antipsychotic drugs (neuroleptics), their ability to block brain dopamine (DA) receptors, and to antagonize psychostimulant-induced stereotypies in animals. For these reasons psychostimulant-induced stereotyped behaviour has long been considered as the most suitable animal model of schizophrenic syndromes.

The interest in using stereotypies for preclinical studies and for screening new neuroleptics has favoured the accumulation of information about the neurobiological basis of these behavioural responses. However, it has also strongly oriented the research and most of the data collected concern the behavioural and biochemical effects of amphetamine, the most widely used psychostimulant, in the rat. Moreover, due to the opposite action exerted by psychostimulants and neuroleptics on DA systems, most of the research on the neurochemistry and neurophysiology of stereotyped behaviour has been conducted on these brain systems. Nonetheless, there are also very recent results concerning the involvement of neuropeptides, especially endogenous opioids, in the modulation of stereotypic behaviour.

More recently, this area of work has received fresh support from data showing that a number of environmental constraints (stress) enhance the ability of psychostimulants to promote stereotypic responses. The phenomenon, known as behavioural sensitization, may be produced also by chronic treatment with psychostimulants. These observations have led some authors to suggest that stress-induced sensitization processes might at least in part be responsible for the development of altered behavioural responses in undrugged organisms. Environmentally induced behavioural sensitization may represent the missing link between research on animal welfare and results drawn from the neurosciences. For this reason, in the following pages I will review literature on pharmacologically induced stereotypic behaviour and on stress-induced sensitization.

Pharmacologically Induced Stereotyped Behaviour

Stereotypic alterations of behavioural sequences have been observed to be produced by a number of pharmacological treatments. The literature on stereotyped responses elicited by psychostimulants, 5-hydroxytryptamine (5-HT) and neuropeptides is reviewed below. This review is not intended to cover all pharmacological treatments capable of inducing stereotypic alterations of behaviour. Instead, an attempt will be made to examine relations between responses elicited by different centrally acting substances in order to seek a general neural hypothesis for stereotypic behaviour.

Psychostimulants

As stated in the Introduction, most neurophysiological studies on stereotypies have looked at amphetamine-induced behaviour in the rat. The precise form or topography of psychostimulant-induced stereotypies varies across species. In the rat it is manifested generally as repetitive sniffing and oral or head movements (Robbins and Sahakian, 1983). This behavioural response is classically observed after treatment with high doses of amphetamine while low to intermediate doses induce increased locomotion. The two are distinct behavioural responses which appear to compete with each other. In fact, the stimulatory effect of amphetamine on gross locomotor activity (the type that can be measured using photocell apparatus) has a characteristic inverted U-shaped evolution depending on the dose (Robbins and Sahakian, 1983), disappearing at high doses when focused stereotypies in almost immobile animals are observed.

By using the neurotoxin 6-hydroxydopamine (6-OHDA), which selectively destroys neurons containing the neurotransmitters dopamine (DA) and noradrenaline (NE), Creese and Iversen (1975) found that the sniffing and head movements induced by high doses of amphetamine were blocked by local injection in the caudatus. There was no reduction, however, in the locomotory response to the drug which, if anything, was enhanced. Furthermore, amphetamine-induced locomotion was prevented in rats receiving local injection of 6-OHDA in a different brain area: the nucleus accumbens. Moreover, since specific depletion of NE did not affect either locomotion or amphetamine-induced stereotypy, it was possible to postulate that two different DA systems modulated stereotypic and locomotor responses to amphetamine.

This conclusion is supported by other studies using similar neurophysiological approaches (see Robbins *et al.*, 1990, for review) and, more recently, by neurochemical investigation using the newly developed techniques of microdialysis. Intracerebral dialysis allows the investigation of regional DA release in awake rats (Ungerstedt, 1984). Using this approach it has been possible to correlate the stereotypic effects of amphetamine with DA release in the caudatus (Sharp et al., 1986), while selective increase of DA release in the nucleus accumbens was found to characterize the central effects of a number of drugs which stimulate locomotion (Di Chiara and Imperato, 1988).

However, precise correlation between the occurrence of stereotypies and the degree of DA release or amphetamine concentration is still to be demonstrated. Instead, contrasting results have been obtained by injecting amphetamine locally in the caudatus. In most cases these experiments were unable to reproduce the full stereotypic repertoire of systemic amphetamine treatment (Robbins et al., 1990). Taken together, these results appear to suggest that stimulation of DA release in the caudatus is a necessary but not sufficient condition for stereotypic behaviour to be produced.

The caudatus putamen is located in the dorsal striatum and receives its DA input from neurons localized in the substantia nigra, making it part of the nigrostriatal DA system. The nucleus accumbens is located in the ventral part of the striatum and receives DA inputs from neurons localized in the ventral tegmental area (VTA), thus it is the projecting area of the mesoaccumbens DA system. DA neurons of the VTA send their axons also toward the frontal cortex (mesocortical DA system) and to the hippocampus and the amygdala, which together with the nucleus accumbens form the mesolimbic DA system. Caudatus and accumbens have largely independent outputs through the dorsal and ventral pallidum respectively. However, because of the neuroanatomical relationships between the two structures (Fig. 6.1), the nucleus accumbens is in a position to alter functioning in the nigrostriatal projection, whereas the reciprocal possibility is absent. Thus the mesoaccumbens and

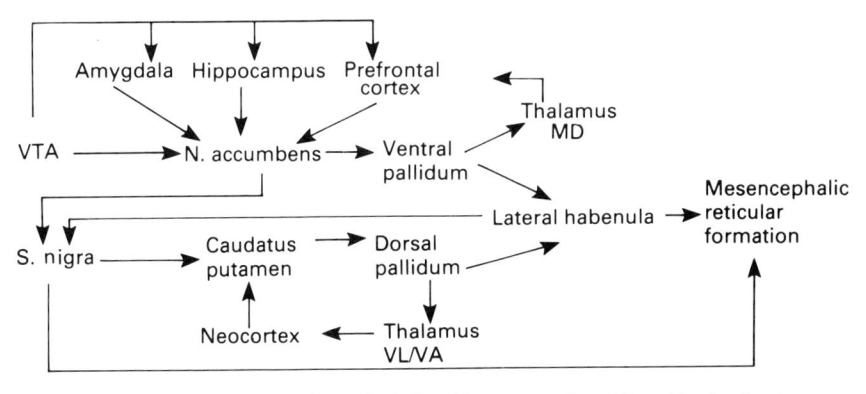

Fig. 6.1. Schematic representations of relationships among the different brain structures that appear to be involved in the control of stereotyped behaviour. The different neurotransmitters implicated are not specified.

nigrostriatal systems have parallel, independent output pathways which may compete, while the accumbens is also capable of interrupting or amplifying nigrostriatal activity.

The fact that neurotoxic injections in the caudatus not only block stereotypic responses but reveal an increase in locomotor activity far greater than that seen at any dose of amphetamine in intact animals supports the view that stereotypic responses induced by the drug in some way compete with the locomotor effect, masking further, dose-dependent increases.

It has been suggested that amphetamine with increasing dosage produces '... an increasing response rate within a reduced number of response categories' (Robbins and Sahakian, 1983) an effect possibly due '... to an underlying general stimulatory effect that was exerted on all responses exceeding some minimal tendency' (Robbins and Sahakian, 1983). DA release in the nucleus accumbens would maintain the general stimulatory effect of the drug and, at low doses of amphetamine, a parallel, moderate DA release from the caudatus would allow the independent, but coordinated stimulation of somewhat distinct response elements permitting quite complex sequences of behaviour to be performed. With increasing doses, the progressively greater activation of the caudatus, which may even be promoted by the nucleus accumbens, occludes the behavioural responses arising from the accumbens which are only apparent when the effects of DA-receptor activation in the caudatus is blocked.

This hypothesis, based on supposed competition between behavioural outputs, implies that administration of amphetamine or other stimulants does not induce deficits in the sensory input and hence that stereotyped behaviour may be under a certain degree of environmental control. This has been demonstrated by a number of results on behavioural effects of amphetamine in different species and in different test conditions (see Robbins *et al.*, 1990, for review). The most striking example of this control is the virtual absence of classical amphetamine-induced stereotypies in animals tested in a novel environment (Sahakian and Robbins, 1975). This result suggests that some brain areas concerned with novelty processing exert an inhibitory action on stereotypies.

In fact, lesions of the hippocampus, a limbic area which sends non-DA outputs to the medial nucleus accumbens, and decortication, which eliminates cortical projection to the dorsal striaum, have been shown to enhance the behavioural effects of amphetamine (Robbins *et al.*, 1990; Whishaw and Mittelman, 1991). Moreover, each type of lesion enhances the specific components of behaviour selectively promoted by the DA area they control. It thus appears that neocortical and limbic influences enable environmental and experiential factors to prevent behaviour overactivated by striatal DA release (both ventral and dorsal) from becoming too divorced from the environmental or historical context of the animal.

Another brain structure that might inhibit DA-dependent stereotypies is the frontal cortex, lesions of which appear to produce behavioural perseveration and oral movements similar to those observed in amphetamine-treated subjects (Ridley and Baker, 1982). Moreover, the destruction of DA fibres projecting from the VTA to the prefrontal cortex produces a behavioural syndrome characterized by locomotor hyperactivity, and loss of the behavioural inhibitory functions responsible for the performance of coordinated tasks (Le Moal et al., 1969; Tassin et al., 1978).

In conclusion, results obtained using different experimental approaches indicate that psychostimulants induce stereotypic behaviour by enhancing DA release at the level of the ventral and dorsal striatum. Moreover, it has been suggested that stereotypy is the culmination of a continuous process of psychomotor stimulation promoted by DA release in the nucleus accumbens, and behavioural competition arising from progressive activation of the nigrostriatal DA system. Finally the two systems are controlled in an inhibitory way by at least three structures (the hippocampus, the neocortex and the frontal cortex), which could play a major role in promoting behavioural variability. Consequently, stereotyped alteration of behaviour may be produced either by overactivation of nigrostriatal and mesoaccumbens DA systems or by reduced activity of one or more of the brain structures that normally control the DA response in the two brain areas.

5-HYDROXYTRYPTAMINE RECEPTOR ACTIVATION

Pharmacological manipulation of postsynaptic 5-HT receptors in rodents leads to complex behavioural syndromes. Their components may include hind limb abduction, 'wet-dog' or body-shakes (paroxysmal shaking of the head and trunk), side-to-side head-weaving, reciprocal forepaw-treading ('piano playing') tremor, Straub tail (rigidly erected tail) and increased reactivity to stimuli (Jacobs, 1976; Gerson and Baldasserini, 1980; Green and Heal, 1985; Curzon, 1990). Head-weaving and forepaw-treading are stereotyped inasmuch as they are apparently meaningless and repetitive fragments of normal behaviour.

Many drugs used to elicit the 5-HT syndromes have other central effects besides increasing 5-HT concentration at receptor level. In particular, at high doses of amphetamine complex behavioural patterns occur which include components of the 5-HT syndrome, classical DA-dependent components and backward walking. Recently, the role of DA in these 5-HT-dependent behavioural responses were investigated by means of local injections of 6-OHDA into the caudatus, the nucleus accumbens, the substantia nigra and the VTA (Andrews et al., 1982)

The results obtained indicate that DA transmission in the caudatus is needed for head-weaving and reciprocal forepaw-treading. These results are in agreement with those showing that haloperidol, the best known DA blocking agent, prevents, and the DA agonist apomorphine potentiates, forepaw treading induced by 5-HT direct agonists. Although contradictory results have been obtained for 5-HT agonist-induced head-weaving, it may generally be concluded that the stereotyped components of 5-HT syndromes (head-weaving and forepaw-treading) are dependent on DA while 5-HT-dependent behaviours without obvious stereotyped character (body-shakes and hindlimb abduction) appear to require neither striatal nor accumbens DA (Curzon, 1990). Indeed, these components tend to increase when central DA levels are reduced, suggesting that they are inhibited by DA transmission (Curzon, 1990).

Finally, although the nature of the interaction between 5-HT and DA systems promoting stereotypies is not clear, recent electrophysiological data seem to suggest that 5-HT systems exert subtle influences on the activity and pharmacological responsiveness of nigrostriatal DA neurons (Kelland et al., 1990).

NEUROPETIDES

Excessive grooming is the typical behavioural response to central injection (intracerebroventricular (i.c.v.) or intracisternal) of a large number of active neuropeptides (Isaacson and Gispen, 1990). This behaviour has been described as more or less repetitive and predictable over an extended time period although it may be interrupted by environmental events. The structure of the behavioural response is not different from environmentally induced grooming (i.e. grooming observed in manipulated animals) or homecage grooming. Thus the most relevant aspect of peptide-induced grooming appears to be its duration, which largely outlasts the response in undrugged animals, and its temporal distribution.

Although a great variety of peptides have been reported to induce excessive grooming, some differences have also been observed in the temporal distribution of the response and in sensitivity to pharmacological manipulations. In the case of ACTH-induced grooming, two phases may be identified: a first phase, which lasts for the first 20 to 30 minutes after i.c.v. injection and is relatively insensitive to most pharmacological agents, and a second phase which may be blocked by systemic neuroleptics or opiate antagonists (Isaacson et al., 1983).

A number of results coming from divergent approaches suggest that the nigrostriatal, as well as part of the mesoaccumbens, DA systems have a strong relation to the mechanisms involved with excessive grooming (Isaacson and Gispen, 1990). Moreover, although grooming is not a component of psychostimulant-induced stereotypies, local injection of amphetamine in specific sites of the dorsal striatum has been shown to elicit a form of grooming together with more classical components of psychostimulant-induced stereotyped movements (Robbins et al., 1990). Finally, some grooming behaviour has been observed in mice and rats that have been fully habituated to the test chambers and injected with DA agonists that selectively activate DA receptors of the D1 type (Molloy and Waddington, 1984).

Although opioid antagonists reduce neuropeptide-induced grooming, and although some opioid peptides and low doses of morphine, may enhance excessive grooming (Isaacson and Gispen, 1990), the response to opiates is mostly behavioural sedation ranging from somnolence to catatonia. Some exceptions occur, particularly in mice, horses, pigs, cats and some humans, in which morphine and enkephalines induce excitement and increased motor activity (see Katz et al., 1978, and Oliverio et al., 1984, for review). In the mouse, in which the phenomenon has been most extensively studied, the motor response to opioids ('running fit' syndrome) has a typical stereotyped nature characterized by almost constant running with a virtual absence of rearing activity, and Straub tail (Katz et al., 1978).

It should be noted that species-specific differences do not completely account for these opposite behavioural effects of opioids. Some strains of mice may exhibit profound behavioural inhibition following morphine injection and respond with a running fit to other types of opioids (Oliverio et al., 1984). These discrepancies have led to the hypothesis that stimulation of different types of opioid receptors may have opposite modulatory effects on behaviour (Oliverio et al., 1984; Longoni et al., 1991). These receptors may have different distribution in different species or strains and may have different sensitivity to different opioid agonists (Oliverio et al., 1984).

Although opiates and opioid peptides modulate dopaminergic neurotransmission in mesoaccumbens, mesocortical and nigrostriatal projections (see Iyengar et al., 1989, and Longoni et al., 1991, for review), the attempts to correlate the behavioural responses to opioids with their modulatory effects on DA release have produced contradictory results. Intra-accumbens injections of opiate produce an initial cataleptic response followed by locomotor stimulation which is resistant to DA antagonists (Pert and Sivit, 1977; Costall et al., 1978; Kalivas et al., 1983). On the other hand, morphine-like opiates stimulate DA release in the accumbens, and opiate-induced hypermotility in mice is selectively blocked by the DA antagonist SCH23390 (Di Chiara and Imperato, 1988; Leone et al., 1991; Longoni et al., 1991).

Once again, these contrasting results may be explained by selective activation of different types of opioid receptors. The motor stimulant effects of systemic morphine appear to be dependent on stimulation of opioid receptors of the μ type in the VTA and in the medial substantia nigra, resulting in disinhibition of DA neurons projecting to the accumbens (Leone et al., 1991). On the other hand, selective stimulation of μ receptors in the nucleus accumbens appears to produce motor inhibition and catalepsy which are independent of DA release. Intra-accumbens injections of (D-Ala^2) Deltorphin II, a natural peptide active on opioid receptors of the δ type, induce strong stereotypic behaviour, identical to that induced by very high doses of amphetamine in rats, which may be selectively blocked by DA D1 receptor blockade (Longoni *et al.*, 1991). Moreover, the δ receptor agonist has a marked stimulatory effect on DA release in the NAS (nucleus accumbens septi). Finally, selective blockade of δ receptors prevents both the behavioural and the neurochemical effects of (D-Ala^2) Deltorphin II (Longoni *et al.*, 1991).

The results suggest that the opioid peptides may exert opposite effects on behaviour depending on their selectivity for subtypes of opioid receptors with different regional distribution in the brain. Moreover, they suggest that opioid-mediated stereotyped behaviour is dependent on the activation of the mesoaccumbens DA system. Finally, the ability of the δ receptor agonist to produce stereotypic behaviour in the absence of striatal DA release suggests that amphetamine-induced behaviour is not the only possible model for understanding the neurophysiology of stereotypies.

Some conclusions can be drawn on the basis of this brief review of pharmacologically induced stereotyped behaviour. First, although different behavioural responses may be selectively elicited in a stereotyped way by different types of central acting substances, increased brain DA activity appears to be a necessary condition for all the different forms of stereotyped behaviour. Second, opioid peptides may exert a major modulatory effect on stereotypies through their different actions on brain, especially accumbens, DA activity. Finally, enhanced DA activity in the caudatus appears to be decisive for the expression of focused stereotypies induced by amphetamine 5-HT receptor agonists and possibly for neuropeptide-induced grooming. However, it is not a requirement for focused stereotypies induced by δ opioid receptor activation.

This last observation indicates a need for caution in evaluating data obtained from the pharmacological manipulation of behaviour before producing general conclusions on the neurophysiology of behaviour in drug-free organisms. One major limit to pharmacological manipulation is the central action of the test substances. Whether they are agonists that mimic the action of an endogenous substance at its receptors or alter turnover of some specific neurotransmitter, thereby increasing its availability, they almost always enhance neurotransmission far beyond physiological levels. Moreover, given their selectivity for specific neurotransmitter or neuromodulator systems, or even for specific types of receptors, they enhance the activity of those brain structures where the highest concentrations of the latter are located, regardless of the complex systems within which they normally function.

One way to overcome these limits is by comparing, wherever possible, the behavioural and neurochemical effects of pharmacological manipulations with the behavioural and neurochemical responses elicited by environmental constraints in drug-free organisms.

Environmentally Induced Stereotypic Behaviour

Various kinds of altered environmental conditions have been reported to produce stereotyped behavioural responses (see Chapters 2 and 3). These kinds of stereotypies are generally described as behavioural responses to stressful or arousing stimuli. However, the interchangeability of the term 'stress' and 'arousal' may be most misleading for the understanding of these phenomena and of their physiological and behavioural effects.

An organism is an open system constantly interacting with its environment (von Bertalanaffy, 1968). When the environment is modified, the organism detects discrepancies between expected and observed events and attempts to cope with the new situation by means of the repertoire of behavioural and physiological responses which are characteristic of its species and of its individual experiences (Wiepkema, 1990). It is this first phase of the organism's response to environmental changes that may be correctly termed 'arousal'. It should be noted that the stimuli which promote this phase are not necessarily aversive or potentially dangerous for the organism's survival (Pribram and McGuinness, 1975). The arousal phase is ended by successful coping with the environmental demand either by means of well-established responses or newly learned ones. However, if coping is prevented by external or internal limitations then a new phase takes over which may be properly called stress (Puglisi-Allegra *et al.*, 1990a,b).

There is some experimental evidence to support the view that stress responses only arise when the organism has lost all its ability to cope with the situation at behavioural level. In their pioneering studies, Weiss and co-workers demonstrated that lack of controllability is the sole condition that produces classic physiological stress responses to aversive situations (see Levine and Wiener, 1989, for review). In these experiments they used rat pairs that received the same amount of shock, but only one animal was able to stop shock delivery by behavioural responses. They showed that only the rat unable to exert behavioural control over the situation exhibited stress responses.

The difference between stressful and arousing conditions may account for contradictory behavioural effects obtained using different experimental 'stressors'. There is good evidence that conditions as diverse as electrical stimulation of the hypothalamus, mild pinching of the tail, and intermittent schedules of reinforcement lead to repetitive oral or locomotor activity which appear similar, although not identical, to behaviour induced by amphetamine (see Robbins *et al.*, 1990, for review). However, several other stressors, such as severe food deprivation, cold, immobilization and exposure to inescapable electric shock, do not seem to lead to stereotyped responses (see Robbins *et al.*, 1990, for review). On the contrary, they produce a typical syndrome characterized by general behavioural depression (Weiss *et al.*, 1981; Anisman and Zacharko, 1990). It is worth noting that this syndrome develops only when behavioural coping or control over the environment is prevented (see Levine and Wiener, 1989, for review); thus it may represent the typical behavioural effect of stress experiences. By contrast, the behavioural activation elicited by some types of stressors represents attempts at behavioural coping during the arousal response. The fact the behaviour observed in these conditions appears to be 'irrelevant' should not be surprising. In this regard, it is interesting to note that controllability over a stressor, which prevents the development of stress responses, can be exerted also by behavioural responses which are unable to affect the stressful stimulus (Weiss et al., 1981).

Although these results shed doubts on the role of stress in the development of environmentally induced stereotypies, there still exists a phenomenon which appears to be generalized among very different types of stressful experiences and that leads to the enhancement of either pharmacologically or environmentally induced stereotypies. This phenomenon is stress-induced behavioural sensitization.

Behavioural sensitization

The term sensitization is generally used to refer to a long-lasting response increment occurring upon repeated presentation of a stimulus that at its initial presentation reliably elicits a response. It is a ubiquitous biological phenomenon that has been used to describe the enhancement of reflexive responses following repeated elicitation, the growth in response to repeated epileptic discharge, and the increased responsiveness of the immune system following initial exposure to an antigen (Groves and Thompson, 1970; Kalivas and Barnes, 1988). For this reason it was argued that sensitization may be a property of cells and one of the simplest forms of memory, which is manifested as a more rapid and stronger response to a stimulus following intermittent exposure to it (Groves and Thompson, 1970; Kandel, 1985).

Repeated systemic injections of different types of psychostimulants lead, among other things, to sensitization of the behavioural-activating effects of these drugs. For example, in animals that have been previously exposed to amphetamine, subsequent amphetamine treatment produces more intense stereotyped behaviour, a reduced time to the onset of stereotypy following injection of the test dose, or the development of stereotyped responses to doses previously devoid of this behavioural effect (Robinson and Becker, 1986; Kalivas and Barnes, 1988). The sensitization to amphetamine does not seem to be a species-specific phenomenon. An enduring behavioural sensitization to repeated intermittent amphetamine administration has been described in rats, mice, cats, guinea-pigs, dogs, non-human primates and humans (see Robinson and Becker, 1986, for review).

It is worth noting, however, that sensitizing processes discriminate between the various elements of the behavioural response to amphetamine. For example, locomotion, stereotyped head and limb movement, and stereotyped sniffing are clearly sensitized by a few injections of amphetamine, but the oral behaviours produced by high doses are not. Moreover, strain-dependent differences in amphetamine-induced sensitization were observed for some but not all the motor effects of the drug in the rats (Robinson and Becker, 1986; Robinson, 1988).

Repeated exposure to mild tail pressure, inescapable footshock, food deprivation and immobilization have all been shown to increase the behavioural responses to amphetamine. Moreover, stereotypic biting induced by tail-pinch is progressively enhanced by repeated exposure either to the stressor or to amphetamine. Finally, rats reared in isolation exhibit both enhanced stereotyped sniffing in response to amphetamine and enhanced tail-pinch induced oral behaviour (see Antelman and Chiodo, 1983, and Robbins et al., 1990, for review). This interchangeability between stress and amphetamine in producing behavioural sensitization has led some authors to suggest that the behavioural effects of amphetamine are due to the stressful properties of this and similar drugs. This hypothesis is supported by the comparative analysis of the neuroendocrine and neuropharmacological effects of stimulants and stressors (Antelman and Chiodo, 1983). However, a number of recent results of research on brain neurochemical responses in stressful conditions have raised strong doubts as to whether stress and psychostimulants have the same central effects.

NEURAL RESPONSES TO STRESS AND PSYCHOSTIMULANTS

As observed above, the brain DA systems appear to play a major role in the expression of the behavioural effects of psychostimulants. The fact that the DA system is also one of the key neurotransmitter systems involved in the central response to stress (see Anisman and Zacharko; 1990, Puglisi-Allegra and Cabib, 1990 and Le Moal and Simon, 1991, for review) seems to support the view that stress and stimulants act upon the same neural mechanisms. However, there are some significant differences between the DA response to stress and stimulants. The first is related to the brain area involved in this response. Several results indicate that stressful experiences preferentially activate the mesocortical and mesoaccumbens DA systems (Lavielle et al., 1979; Deutch et al., 1985; Imperato et al., 1989; Puglisi-Allegra and Cabib, 1990). Evidence for the involvement of the nigrostriatal system in central responses to stress is far less consistent (Dunn and File, 1983; Cabib et al., 1988a, b; Abercrombie et al., 1989). The mesocortical DA system seems to be especially sensitive to stressful stimuli since it is activated also by very mild and brief exposure to stressors. If, as proposed earlier, this system inhibits the subcortical DA projections, then it may represent a mechanism for regulating stress responses and preventing the deleterious behavioural effects of overactivation. A recent result showing anticipation of mesoaccumbens DA response to stress in animals bearing frontal lesions appears to support this view (Deutch et al., 1990).

The absence of a nigrostriatal DA response in several types of commonly used stressful procedures would explain the absence of stereotyped responses in animals acutely exposed to these types of stressors. In this regard, it is worth noting that stereotyped biting induced by tail-pinch, is absent in rats treated with intra-striatal 6-OHDA (Antelman and Szechtam, 1975), which possibly suggests that some specific stressors may also activate the nigrostriatal system. These conclusions should be viewed with caution. Endogenous opioids are massively released during stressful experiences (Amir et al., 1981), and have been shown to exert modulatory effects on mesoaccumbens and mesocortical DA responses to stress (Miller et al., 1984; Cabib et al., 1989) and to produce stereotyped responses which are independent of nigrostriatal DA activation (Longoni et al., 1991). Thus stereotypic responses induced by some stressors may be due to opioid-dependent activation of mesoaccumbens DA.

Finally, the absence of a striatal response to acute stress does not rule out the possibility that repeated exposure to stressful experience leads to adaptive changes at the level of the striatum given the postulated relationship between mesoaccumbens and nigrostriatal structures. In fact, enhanced amphetamine-stimulated striatal DA release following prior stress exposure has been observed *in vitro* (Robinson, 1988). Nonetheless, it was shown that rats with extensive, bilateral DA-depleting lesions of the caudate putamen develop behavioural sensitization and parallel reduction of corticosterone response following chronic amphetamine treatment (Mittelman et al., 1991). These findings suggest that adaptive processes leading to sensitization do not operate at the level of the nigrostriatal system. Moreover, cross-sensitization between the locomotor activating effects of amphetamine and morphine or intra-VTA enkephaline has been reported (Vezina and Stewart, 1989) which suggests that opioids, like psychostimulants, produce behavioural sensitization possibly through a similar brain mechanism. Finally, local injection of DA antagonists in the VTA prevents the development of both amphetamine- and opioid-induced behavioural sensitization indicating that sensitizing processes involve presynaptic mechanisms of the mesoaccumbens or mesocortical DA systems.

The other difference between the brain DA response to stress and psychostimulants relates to the quality of this response. By examining DA release in the nucleus accumbens using the microdialysis method, Di Chiara and Imperato (1988) showed that stimulants induce an increase in DA release in this brain structure lasting for some hours after injection. Given the half life of these substances in the brain, it may be concluded that their effect is indeed to increase DA release. By contrast, using the same method, Puglisi-Allegra et al. (1991) have shown that rats subjected to restraint show an increase in DA release in the nucleus accumbens lasting up to 50 minutes, followed by a

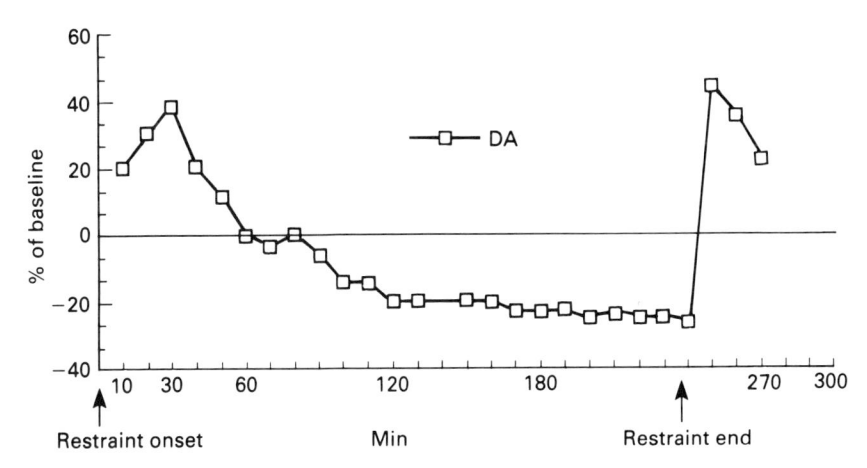

Fig. 6.2. Effects of restraint stress on DA release in the nucleus accumbens as measured by intracerebral dialysis in rats. Results are expressed as percentage changes from basal levels.

decrease below control levels lasting as long as the stressful condition is maintained (Fig. 6.2). The biphasic evolution of mesoaccumbens DA response to stress has also been described in mice exposed to either restraint or footshock stress using *ex vivo* methods of neurochemical analysis (Puglisi-Allegra *et al.*, 1991). It thus appears that, although both psychostimulants and stress induce a response in the mesoaccumbens DA system, the time-dependent evolution of this response differs between the two treatments. The opposite changes in DA release during stress exposure may be related to the passage from an earlier arousal phase characterized by DA activation to a later stress condition which leads to inhibition of DA release and further emphasizes the difference between the two phases.

Once again, the difference in the response to a single stress or psychostimulant experience does not necessarily rule out the possibility that adaptation to repeated exposure produces similar outcomes. Several results suggest that disruption of some presynaptic DA processes may interfere with the development of behavioural sensitization induced by repeated amphetamine (see Robinson, 1988, for review). It should be noted that DA is not exclusively released from axons in the synapses of the terminal field. The neurotransmitter is also released by the cell bodies and dendrites of the DA neurones (Fig. 6.3) in the substantia nigra in the case of the nigrostriatal systems, and in the VTA in the case of mesolimbic system. DA release in these areas may not follow the same evolution as the release in the terminal fields (Nieoullon *et al.*, 1978).

Moreover, sensitization may well involve non-dopaminergic systems. For example, the activation of the hypothalamo–pituitary–adrenocortical axis, the most classic physiological response to stress, has been shown to be

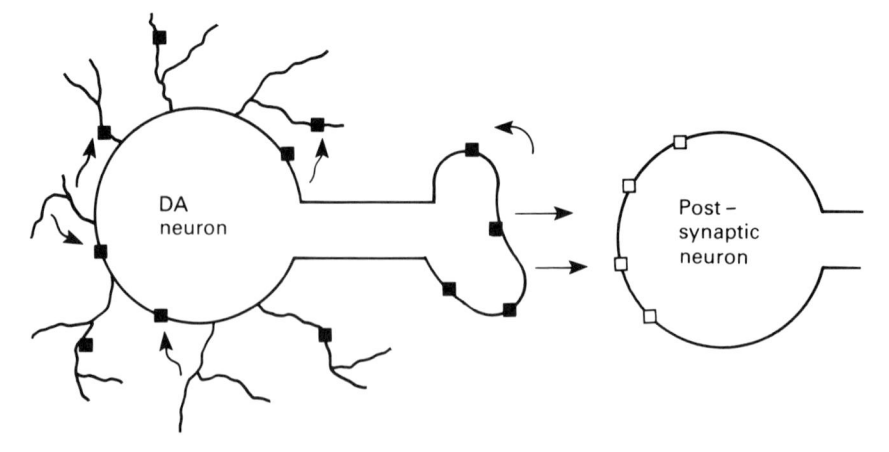

Fig. 6.3. Schematic representation of a DA neuron showing pre- (filled-in squares) and postsynaptic (open squares) receptors. Arrows indicate DA release.

necessary for the development of amphetamine-induced sensitization (Rivet et al., 1989). Although recent results appear to shed doubts on a direct role of corticosterone in the modulation of brain DA release during stress (Imperato et al., 1991), direct interaction between glucocorticoids and DA receptors may be responsible for changes in receptor sensitivity to the neurotransmitter, as already observed for different neurotransmitters (e.g. Stone, 1990), leading to altered functioning of the DA system.

In conclusion, the hypothesis that stress and psychostimulants represent the same process does not appear to be supported by experimental data. Moreover, the data reviewed suggest that chronic or repeated exposure to stressors may well produce sensitization to stimulants by inducing major alterations in brain DA functioning.

Behavioural Sensitization and Adaptation to Environmental Constraints

In the preceding pages the hypothesis was proposed that stress-induced behavioural sensitization may be dependent on altered brain DA functioning. These alterations may be the result of adaptation to chronic or repeated stressful experiences. It is worth noting in this regard that MacLennan and Maier (1983) have shown that a regimen of severe electric shocks enhances the behavioural response to amphetamine, but only if the shocks are uncontrollable. Exposure to uncontrollable electric shocks is the typical procedure used to induce behavioural depression in rats, a phenomenon that is progressively reduced following repeated exposure to the stressful stimuli. This last observation may suggest that adaptive processes are activated to overcome the impairing effects of stress. Recent results have pointed to an inverse relationship between stereotypic responses and levels of plasma corticosterone in amphetamine-treated rats. The increase in plasma corticosterone induced by amphetamine was, in fact, enhanced by caudate lesions which reduce the amount of stereotyped behaviour elicited by the psychostimulant and reduced in animals exhibiting behavioural sensitization (Mittelman et al., 1991). These results support the view that increased stereotypy may be produced by coping processes.

Some of the brain adaptation processes to repeated stress involve NE-dependent systems and are based on changes in receptor sensitivity to the transmitter (Stone, 1990). Moreover, alterations of 5-HT and the endogenous opioid system functioning have been observed in chronically stressed animals (Amir et al., 1981; Molina et al., 1990). However, in view of the role played by the mesoaccumbens DA system in the development and in the expression of behavioural sensitization, special attention should be paid to the adaptation processes that involve this brain system.

In the following pages, I shall review experimental evidence supporting the view that behavioural sensitization may result from adaptation to repeated stress involving the mesoaccumbens DA system and it will be suggested that the neural mechanism involved in the development of behavioural sensitization may also be responsible for the appearance of stereotypic responses in undrugged organisms.

BEHAVIOURAL ADAPTATION TO STRESS: ROLE OF GENOTYPE AND OF THE MESOACCUMBENS DOPAMINE SYSTEM

Large individual differences have been observed in different species for the effects of repeated amphetamine treatment as well as for sensitivity to the behavioural effect of a first injection of the psychostimulants. The data collected clearly indicate that the initial response to the drug does not predict individual vulnerability to amphetamine-induced behavioural sensitization (Robinson, 1988). However, the existence of robust strain differences in the effect of chronic amphetamine treatment (Robinson, 1988) suggests that genetic factors influence behavioural sensitization processes. Consequently, the study of genotype-dependent differences in brain and behavioural adaptation to repeated stressful experience may shed some light on the neural basis of environmentally induced stereotypies.

The climbing response is a widely used behavioural probe to test brain DA functioning (Martres et al., 1977; Costall et al., 1979). Climbing or clinging to the wire mesh of the cage cover is a behaviour easily observed when mice are in their breeding cages, and to a greater extent when mice are attacked by the dominant male or disturbed by a researcher's manipulations. Moreover, mice exposed to novelty show a marked increase in locomotor behaviour, defecation, plasma corticosterone levels and mesoaccumbens DA metabolism. It has been suggested that these responses are elicited by the test situation, usually a brightly illuminated large enclosed space, which represents danger for rodents, and that the heightened activity is produced by the attempts made to escape or to find a shelter (Misslin et al., 1982; Cabib et al., 1990). Since it has been shown that in these situations mice also show a vigorous climbing response (Cabib et al., 1990), this behaviour may be a defensive response to potentially dangerous situations.

During prolonged exposure to the novel environment, the amount of climbing behaviour decreases progressively while the mice become more or less immobile (Cabib et al., 1990). Although habituation may be an economic explanation for the progressive reduction of climbing behaviour, an alternative possibility could be that a stress syndrome follows the initial arousal reaction producing behavioural depression. In fact, it is conceivable that prolonged exposure to an aversive environment devoid of ways to escape represents for the animal failure to cope resulting in stress.

This conclusion is supported by data showing that previous experiences of stress markedly affect climbing behaviour elicited by a novel environment. We have observed that prolonged exposure to stress (120 minutes of restraint) decreases climbing behaviour in mice (Cabib et al., 1988a). Since this behavioural response is modulated by the brain dopamine system (Martres et al., 1977; Costall et al., 1979), the inhibitory effect of stress on climbing may be tentatively related to stress-induced inhibition of DA activity in the nucleus accumbens (Cabib et al., 1988b). However, while the behavioural inhibition produced by previous stress exposure disappears after repeated stress experiences, thus indicating adaptation, no changes in the effects of stress on mesoaccumbens DA release are observable following repeated stress exposure (Puglisi-Allegra et al., 1990b).

Adaptation of the climbing response to repeated stress experiences reveals marked strain differences. Repeated exposure to restraint prevents the behavioural effects of acute stress in mice of the DBA/2 (DBA) strain but dramatically enhances them in mice of the C57BL/6 (C57) strain. However, while the similar behavioural effects of acute stress in the two strains are in accordance with similar inhibitory effects on mesoaccumbens DA release, neither strain shows adaptation to repeated stress as far as the neurochemical response is concerned.

These results suggest that adaptation to repeated stress experiences involves the alteration of DA-receptor response to the neurotransmitter. Brain DA receptors are extremely plastic and their sensitivity may be either increased or decreased by a number of experimental manipulations (see Martres et al., 1977, for review). Following repeated stress, DBA mice show a reduced sensitivity to the ability of low doses of the DA agonist apomorphine to inhibit both climbing behaviour and mesoaccumbens DA release. By contrast, a marked increase of both the behavioural and the biochemical effects of these same doses of apomorphine were observed in repeatedly stressed C57 mice. A genetic analysis involving F1 and F2 hybrids and the two backcrossed populations indicated a significant interaction between genotype and stress in the expression of these phenotypes and a dominant inheritance of the C57 response to repeated stress (Cabib et al., 1984; Cabib and Puglisi-Allegra, 1991).

The low doses of apomorphine used for these experiments selectively activate DA autoreceptors (see Fig. 6.3). Consequently, a reduction of effects of low doses of apomorphine indicates decreased sensitivity while an enhanced effect of the agonist indicates sensitization of these receptors. Moreover, DA autoreceptors are localized on the DA cells and have an inhibitory effect on cell activity and it has been suggested that they represent a feedback mechanism regulating DA transmission (see Martres et al., 1977, for review). Thus altered sensitivity of this kind of receptor leads to altered functioning of one of the regulatory mechanisms of the DA systems.

Reduced sensitivity of DA autoreceptors in the mesoaccumbens system of repeatedly stressed DBA mice may represent adaptation to stress-induced inhibition of DA release in this area. The adaptive alteration of brain functioning, in turn, would allow the maintenance of a defensive response, climbing, in conditions of chronic stress. On the other hand, repeatedly stressed DBA mice show increased sensitivity to the behavioural effects of amphetamine (Badiani et al., 1992), the typical symptom of behavioural sensitization, that may also be dependent on the reduced activity of the regulatory mechanism based on DA autoreceptors. Indirect support for this hypothesis comes from the results obtained with mice of the C57 strain and the F1 hybrids. Following repeated stress experiences, in fact, both C57 and F1 hybrids show reduced sensitivity to the behavioural effects of amphetamine. Since alteration of DA autoreceptor sensitivity characteristic of C57 mice is inherited in a dominant way in crossed populations (Cabib et al., 1984), the observation that repeatedly stressed F1 hybrids present the same behavioural response to amphetamine as their C57 parental strain strongly suggests the involvement of changes in autoreceptor sensitivity in strain-dependent sensitization to the behavioural effects of amphetamine (Fig. 6.4).

It should be noted that unstressed C57 mice are far more sensitive to the stimulatory effects of amphetamine on locomotion while stress-induced behavioural sensitization produces a complete inversion of this strain difference leading to DBA mice being more sensitive than C57. These results, in agreement with those obtained in a different species (Terlouw et al., 1992), support the hypothesis that vulnerability to stress-induced sensitization does not depend on the initial response to the drug.

Taken together, these results support the view that, depending on genotype, repeated exposure to stressful stimuli may induce adaptation processes involving reduced ability of the mesoaccumbens DA system to control its own activity through the autoreceptor feedback. This leads to a state of

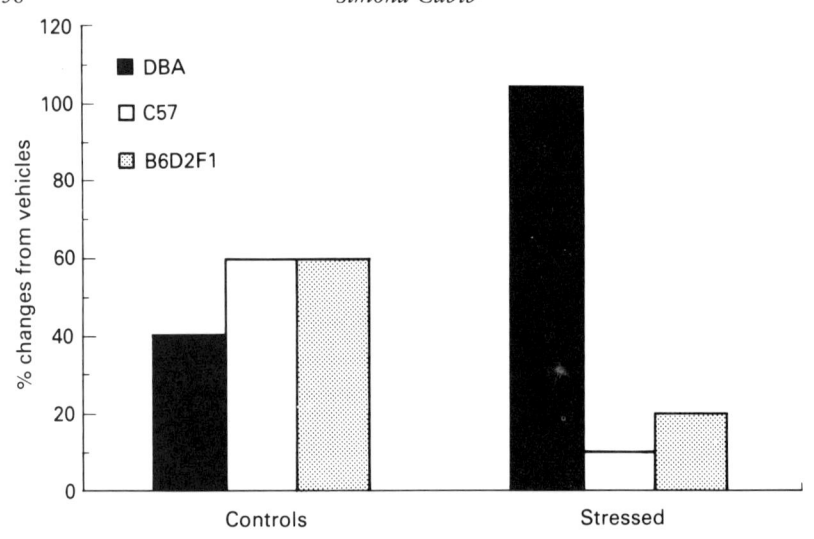

Fig. 6.4. Effects of ten daily sessions of restraint stress on amphetamine-induced locomotion in two inbred strains of mice and in their F1 hybrids tested 24 hours after the last stressful experience.

hyperresponsivity of the system to pharmacological challenge: i.e. behavioural sensitization.

ADAPTATION TO STRESS AND BEHAVIOURAL DYSFUNCTIONS

As previously observed, behavioural sensitization may be produced by virtually all kinds of manipulations leading either to repeated or chronic exposure to stressful conditions. Moreover, several results suggest that disruption of some presynaptic DA processes may interfere with the development of behavioural sensitization induced by repeated amphetamine (see Robinson, 1988, for review). Thus, mesoaccumbens DA hyperactivity may represent a general neurological substrate of behavioural sensitization (Eichler and Antelman, 1979).

It is worth noting at this point, that a state of hyperactivity of the DA mesoaccumbens system may markedly alter the general organism–environment interaction.

As already observed, increased DA release is the characteristic arousal response of the mesoaccumbens system. The arousal response is elicited by any change in the environment. Thus the mesoaccumbens DA system may be activated during the first phase of response to stressful events but also by non-aversive or even pleasurable stimuli (Wise, 1980; Simon and Le Moal, 1984; Le Moal and Simon, 1991; Puglisi-Allegra et al., 1991). Furthermore, it has been shown that the mesoaccumbens DA system is not only activated by arousing situations but also by neutral stimuli previously paired with aversive experiences (Herman et al., 1982; Puglisi-Allegra and Cabib, 1990). It is thus conceivable that a mesoaccumbens DA system adapted to stress would be hyperresponsive to many different stimuli.

From a behavioural point of view, mesoaccumbens DA hyperactivity may well promote stereotypies. We have previously observed that activation of the mesoaccumbens DA system is necessary for the expression of amphetamine-induced stereotypic behaviour and both necessary and sufficient for opioid-induced stereotypies. Consequently, in arousing situations (ranging from very mild noxious stimuli to the expectation of food delivery), the hyperactive DA mesoaccumbens system might alter the behavioural outputs of the organism in a stereotypic way.

Moreover, it has been shown that 'displacement behaviours', those apparently 'irrelevant' activities which often intrude, either during conflict tendencies, or when a strong tendency is thwarted, are dependent on the integrity of the mesoaccumbens DA projections (Robbins and Koob, 1980). This is not surprising in view of the previously discussed sensitivity of the system to arousal, one of the major factors involved in the promotion of this kind of behaviour (Hinde, 1970). In chronically stressed organisms, displacement behaviours may become overwhelming due to sensitization of the mesoaccumbens system to arousing stimuli, finally leading to bouts of highly stereotyped behaviour. This conclusion is supported by the observation that most pharmacologically induced stereotypies observed in the rat (grooming, sniffing, gnawing) also appear as displacement activities in drug-free animals (Hinde, 1970). Furthermore, these activities often appear as incomplete sequences of behaviour (Hinde, 1970). Thus overstimulation could increase fragmentation of behavioural sequences leading to '... an increasing response rate within a reduced number of response categories' (Robbins and Sahakian, 1983) characteristic of pharmacologically induced stereotypies.

The importance of the role played by the mesoaccumbens DA system in stress-induced sensitization and in possible behavioural alterations produced by this phenomenon should not be interpreted as exclusive. As previously observed, the mesoaccumbens DA system involves complex interactions among limbic and cortical structures which also receive inputs from DA neurons of the VTA. It is interesting to note that in rats bearing hippocampal lesions the signalled presentation of reward (food following deprivation) is sufficient to produce behaviour resembling amphetamine-induced stereotypies and locomotion (Devenport et al., 1981). Moreover, it has been observed that lesions of the fimbria fornix, a hippocampal-accumbens pathway, can block the development of amphetamine-induced behavioural sensitization (Yoshikawa et al., 1991). Finally, since the nucleus accumbens may directly modulate nigrostriatal activity it could also lead to potentiation of those responses promoted by stressors which selectively activate the nigrostriatal DA system (as, for example, tail-pinch-induced oral behaviour).

Moreover, corticosterone can influence receptor sensitivity and is necessary for the development of behavioural sensitization (Rivet et al., 1989; Stone, 1990). Interaction between glucocorticoids and mesoaccumbens DA autoreceptors may represent a regulatory mechanism producing both adaptation to stress and susceptibility to behavioural sensitization (Mittelman et al., 1991). Finally, the opioid systems have been shown to play a major role in the modulation of mesoaccumbens and mesocortical DA responses to stress (Miller et al., 1984; Cabib et al., 1989), and in the development of alteration of DA-mediated behaviour in repeatedly stressed mice (Cabib et al., 1985, 1989). Furthermore, as described above, opioid systems exert a strong modulatory action on behaviour as well as on the activity of the mesoaccumbens DA system. Taken together, these results indicate that non-DA systems may play a major role in the development of environmentally induced stereotypies as well as in their expression and/or maintenance.

In the previous pages, I have reviewed data from research on pharmacologically induced stereotypies and on stress-induced behavioural sensitization to put forward an hypothesis on the neurobiological basis of stereotyped behaviour in drug-free animals. This hypothesis suggests that adaptation to chronic or repeated stress experiences may produce sensitization of the mesoaccumbens DA system to arousing stimuli leading to altered behavioural output. It has been suggested that stereotyped behaviour may serve a coping function to modulate stress responses elicited by environmental stimuli or psychostimulants (Valenstein, 1976; Mittelman et al., 1991). The hypothesis presented here is only partially in line with this view. In fact, it suggests that displacement behaviours, although apparently irrelevant, may protect the organism from entering a stress phase in inescapable or otherwise uncontrollable situations. However, it also suggests that stereotyped alteration of behaviour may be a 'side-effect' induced by the physiological changes involved in adaptation to repeated or chronic stress conditions. This last conclusion is supported by the observation that genetic factors play a major role in promoting stress-induced sensitization and related behavioural alterations.

Consequently, behavioural sensitization does not represent a necessary outcome of adaptation to environmental constraints.

References

Abercrombie, E.D., Keefe, K.A., DiFrischia, D.F. and Zigmond, M.J. (1989) Differential effects of stress on *in vivo* dopamine release in striatum, nucleus accumbens and medial frontal cortex. *Journal of Neurochemistry* 52, 1655–1658.

Amir, S.Z., Brown, Z.W. and Amit, Z. (1981) The role of endorphines in stress: Evidence and speculations. *Neuroscience Biobehavioural Review* 4, 77–86.

Andrews, C.D., Fernando, J.C.R. and Curzon, G. (1982) Differential involvement of

dopamine-containing tracts in 5-hydroxytryptamine-dependent behaviours caused by amphetamine in large doses. *Neuropharmacology* 21, 63–68.
Anisman, H and Zacharko, R.M. (1990) Multiple neurochemical and behavioural consequences of stressors: implications for depression. *Pharmacology and Therapeutics* 46, 119–136.
Antelman, S.M. and Chiodo, L.S. (1983) Amphetamine as a stressor. In: Creese, I. (ed.), *Stimulants: Neurochemical, Behavioural and Clinical Perspectives.* Raven Press, New York, pp. 269–299.
Antelman, S.M. and Szechtam, H. (1975) Tail pinch induces eating in sated rats which appear to depend on nigrostriatal dopamine. *Science* 189, 731–733.
Badiani, A., Cabib, S. and Puglisi-Allegra, S. (1992) Chronic stress induces strain-dependent sensitization to the behavioural effects of amphetamine in the mouse. *Pharmacology, Biochemistry and Behaviour* 43, 53–60.
Bertalanaffy, L. von (1968) *General System Theory, Essays on its Foundation and Development.* Braziller, New York.
Cabib, S. and Puglisi-Allegra, S. (1991) Genotype-dependent effects of chronic stress on striatal and mesolimbic dopamine metabolism in response to apomorphine. *Brain Research* 542, 91–96.
Cabib, S., Puglisi-Allegra, S. and Oliverio, A. (1984) Chronic stress enhances apomorphine-induced climbing behaviour in mice: role of endogenous opioids. *Brain Research* 298, 138–140.
Cabib, S., Puglisi-Allegra, S. and Oliverio, A. (1985) A genetic analysis of stereotypy in the mouse. *Behavioural and Neural Biology* 44, 239–248.
Cabib, S., Kempf, E., Schleef, C., Mele, A. and Puglisi-Allegra, S. (1988a) Different effects of acute and chronic stress in two dopamine mediated behaviours in the mouse. *Physiology and Behaviour* 43, 223–227.
Cabib, S., Kempf, S., Schleef, C., Oliverio, A. and Puglisi-Allegra, S. (1988b) Effects of immobilization stress on dopamine and its metabolites in different brain areas of the mouse: role of genotype and stress duration. *Brain Research* 441, 153–160.
Cabib, S., Oliverio, A. and Puglisi-Allegra, S. (1989) Stress-induced decrease of 3-methoxytyramine in the nucleus accumbens of the mouse is prevented by naltrexone pretreatment. *Life Sciences* 45, 1031–1037.
Cabib, S., Algeri, S., Perego, C. and Puglisi-Allegra, S. (1990) Behavioural and biochemical changes monitored in two inbred strains of mice during exploration of an unfamiliar environment. *Physiology and Behaviour* 47, 749–753.
Costall, B., Fortune, D.H. and Naylor, R.J. (1978) The induction of catalepsy and hyperactivity by morphine administered directly in the nucleus accumbens of rats. *European Journal of Pharmacology* 49, 49–64.
Costal, B., Naylor, R.J. and Nohria, V. (1979) Hyperactivity response to apomorphine and amphetamine in the mouse: the importance of the nucleus accumbens and caudate putamen. *Journal of Pharmacy and Pharmacology* 31, 259–261.
Creese, I. and Iversen, S.D. (1975) The pharmacological and anatomical substrates of the amphetamine response in the rat. *Brain Research* 83, 419–436.
Curzon, G. (1990) Stereotyped and other motor responses of 5-hydroxytryptamine receptor activation. In: Cooper, S. J. and Dourish, C. T. (eds), *Neurobiology of Stereotyped Behaviour.* Clarendon Press, Oxford, pp. 142–168.
Dantzer, R. (1986) Behavioural, physiological and functional aspects of stereotyped behaviour: a review and a reinterpretation. *Journal of Animal Science* 62, 1776–1786.

Deutch, A. Y., Tam, S. and Roth, R. (1985) Foot-shock and conditioned stress increase 3,4-dihydroxyphenilacetic acid (DOPAC) in the ventral tegmental area but not in the substantia nigra. *Brain Research* 33, 143–146.

Deutch, A.Y., Clark, A. and Roth, R.H. (1990) Prefrontal cortical dopamine depletion enhances the responsiveness of mesolimbic dopamine neurons to stress. *Brain Research* 521, 311–315.

Devenport, L.D., Devenport, J.A. and Holloway, F.A. (1981) Stereotypy: modulation by the accumbens. *Science* 212, 1288–1289.

Di Chiara, G. and Imperato, A. (1988) Drugs abused by humans preferentially increase synaptic dopamine concentration in the mesolimbic system of freely moving rat. *Proceedings of the National Academy of Science, USA* 85, 5274–5278.

Dunn, A.J. and File, S.E. (1983) Cold restraint alters dopamine metabolism in frontal cortex, nucleus accumbens and neostriatum. *Physiology and Behaviour* 31, 511–513.

Eichler, A.J. and Antelman, S.M. (1979) Sensitization to amphetamine and stress may involve nucleus accumbens and medial frontal cortex. *Brain Research* 176, 412–416.

Gerson, S. C. and Baldasserini, R. J. (1980) Motor effects of serotonin in the central nervous system. *Life Sciences* 27, 1435–1451.

Green, A.R. and Heal, D.J. (1985) The effects of drugs on serotonin mediated behavioural models. In: Green, A.R. (ed.), *Neuropharmacology of Serotonin*. Oxford University Press, Oxford, pp. 326–365.

Groves, P.M. and Thompson, R.F. (1970) Habituation: a dual process theory. *Psychological Review* 77, 419–450.

Herman, J.P., Guillenau, D., Dantzer, R., Scatton, B., Semerdjian-Roquier, L. and Le Moal, M. (1982) Differential effects of inescapable footshocks and of stimuli previously paired with inescapable footshocks on dopamine turnover in cortical and limbic areas of the rat. *Life Sciences* 30, 2207–2214.

Hinde, R.A. (1970) *Animal Behaviour: A Synthesis of Ethology and Comparative Psychology*. McGraw-Hill, New York.

Imperato, A., Puglisi-Allegra, S., Casolini, P., Zocchi, A. and Angelucci, L. (1989) Stress-induced enhancement of dopamine and acetylcholine release in limbic structures: role of corticosterone. *European Journal of Pharmacology* 165, 337–338.

Imperato, A., Puglisi-Allegra, S., Casolini, P. and Angelucci, L. (1991) Changes in dopamine and acetylcholine release during and following stress are independent of the pituitary–adrenocortical axis. *Brain Research* 538, 111–117.

Isaacson, R.L. and Gispen, W.H. (1990) Neuropeptides and the issue of stereotypy in behaviour. In: Cooper, S.J. and Dourish, C.T. (eds), *Neurobiology of Stereotyped Behaviour*. Clarendon Press, Oxford, pp. 118–141.

Isaacson, R.L., Hannigan, J.H. and Gispen, W.H. (1983) The time course of excessive grooming after neuropeptide administration. *Brain Research Bulletin* 11, 289–293.

Iyengar, S., Kim, H.S., Marien, M.R., McHugh, D. and Wood, P.L. (1989) Modulation of mesolimbic dopaminergic projection by beta-endorphine in the rat. *Neuropharmacology* 28, 123–128.

Jacobs, B.L. (1976) Mini review. An animal model for studying central serotoninergic synapses. *Life Sciences* 19, 777–786.

Kalivas, P.W. and Barnes, C. (1988) *Sensitization of The Nervous System*. Telford Press, Caldwell, New Jersey.

Kalivas, P.W., Wilderlow, E., Stanley, D., Bresse, G. and Prange, A.J. (1983) Enkephaline action in the mesolimbic system: a dopamine dependent and a dopamine independent increase of locomotor activity. *Journal of Pharmacology and Experimental Therapeutics* 227, 229–237.

Kandel, E.R. (1985) Cellular mechanism of learning and the biological basis of individuality. In: Kandel, E.R. and Shwartz J.H. (eds), *Principles of Neurosciences*. Elsevier, New York, pp. 817–833.

Katz, R.J., Carrol, B.J. and Baldrighi, J. (1978) Behavioural activation by enkephalins in mice. *Pharmacology, Biochemistry and Behaviour* 8, 493–496.

Kelland, M.D., Freeman, A.S. and Chiodo, L.A. (1990) Serotoninergic afferent regulation of the basic physiology and pharmacological responsiveness of nigrostriatal dopamine neurons. *Journal of Pharmacology and Experimental Therapeutics* 253, 803–806.

Lavielle, S., Tassin, J.P., Thierry, A.M., Blanc, G.G., Herve, D., Barthelamy, C. and Glowinski, J. (1979) Blockade by benzodiazepines of the selective increase in dopamine turnover induced by stress in mesocortical dopaminergic neurons of the rat. *Brain Research* 168, 585–594.

Le Moal, M. and Simon, H. (1991) Mesocorticolimbic dopaminergic network: functional and regulatory role. *Physiological Reviews* 71, 155–234.

Le Moal, M., Cardo, B. and Stinus, L. (1969) Influence of ventral mesencephalic lesions on various spontaneous and conditioned behaviours in the rat. *Physiology and Behaviour* 4, 567–574.

Leone, P., Pocock, D. and Wise, R.A. (1991) Morphine–dopamine interaction: ventral tegmental morphine and nucleus accumbens dopamine release. *Pharmacology, Biochemistry and Behaviour* 39, 469–472.

Levine, S. and Wiener, S.G. (1989) Coping with uncertainty: a paradox. In: Palermo, D.S. (eds), *Coping with Uncertainty: Behavioural and Developmental Perspectives*. Lawrence Erlbaum Associates, Hillsdale, New Jersey pp. 1–16.

Longoni, R., Spina, L., Mulas, A., Carboni, E., Garau, L., Melchiorri, P. and Di Chiara, G. (1991) (D-Ala^2)Deltorphine II: D1-dependent stereotypies and stimulation of dopamine release in the nucleus accumbens. *Journal of Neuroscience* 11, 1565–1576.

MacLennan, A.J. and Maier, S.F. (1983) Coping and stress-induced potentiation of stimulant stereotypy in the rat. *Science* 219, 1091–1093.

Martres, M.P., Constentin, J., Baudry, M., Marcais, H., Protais, P. and Schwartz, J.C. (1977) Long-term changes in sensitivity of pre- and postsynaptic dopamine receptors in the mouse striatum evidenced by behavioural and biochemical studies. *Brain Research* 136, 319–337.

Miller, J.D., Speciale, S.G., McMillen, B.A. and German, D.C. (1984) Naloxone antagonism of stress-induced augmentation of frontal cortex dopamine metabolism. *European Journal of Pharmacology* 98, 437–439.

Misslin, R., Herzog, F., Koch, B. and Ropartz, P. (1982) Effects of isolation, handling and novelty on pituitary–adrenal response in the mouse. *Psychoneuroendocrinology* 7, 217–221.

Mittelman, G., Jones, G.H. and Robbins, T.W. (1991) Sensitization of amphetamine-stereotypy reduces plasma corticosterone: implications for stereotypy as a coping response. *Behavioural and Neural Biology* 56, 170–182.

Molina, V.A., Volosin, M., Cancela, L., Keller, E., Murua, V.S. and Basso, M. (1990) Effect of chronic stress on monoamine receptors: influence of imipramine administration. *Pharmacology, Biochemistry and Behaviour* 35, 335–340.

Molloy, A.G. and Waddington, J.L. (1984) Dopaminergic behaviour stereospecifically promoted by D-1 agonist R-SKF 38393 and selectively blocked by the D-1 antagonist SCH 23390. *Psychopharmacology* 82, 409–410.

Nieoullon, A., Cheramy, A. and Glowinski, J. (1978) Release of dopamine in both substantia nigrae in response to unilateral stimulation of cerebellar nuclei of cat. *Brain Research* 148, 143–152.

Oliverio, A., Castellano, C. and Puglisi-Allegra, S. (1984) Psychobiology of opioids. *International Review of Neurobiology* 25, 277–337.

Pert, A. and Sivit, C. (1977) Neuroanatomical focus for morphine and enkephaline-induced hypermotility. *Nature* 265, 645–647.

Pribram, K.H. and McGuinness, D. (1975) Arousal, activation and effort in the control of attention. *Psychological Review* 82, 116–149.

Puglisi-Allegra, S. and Cabib, S. (1990) Effects of defeat experiences on dopamine metabolism in different brain areas of the mouse. *Aggressive Behaviour* 16, 271–283.

Puglisi-Allegra, S., Cabib, S., Kempf, E. and Oliverio, A. (1990a) Genotype-dependent adaptation of brain dopamine system to stress. In: Puglisi-Allegra, S. and Oliverio, A. (eds), *Psychobiology of Stress*. Kluwer Academic Press, Dordrecht, pp. 171–182.

Puglisi-Allegra, S., Kempf, E. and Cabib, S. (1990b) Role of genotype in the adaptation of the brain dopamine system to stress. *Neuroscience Biobehavioural Review* 14, 523–528.

Puglisi-Allegra, S., Imperato, A., Angelucci, L. and Cabib, S. (1991) Acute stress induces time-dependent responses in dopamine mesolimbic system. *Brain Research* 554, 217–222.

Ridley, R.M. and Baker, H.F. (1982) Stereotypies in monkeys and humans. *Psychological Medicine* 12, 61–72.

Rivet, J.-M., Stinus, L., Le Moal, M. and Morméde, P. (1989) Behavioural sensitization to amphetamine is dependent on corticosteroid receptor activation. *Brain Research* 498, 149–153.

Robbins, T.W. and Koob, G.F. (1980) Selective disruption of displacement behaviour by lesions of the mesolimbic dopamine system. *Nature* 285, 409–412.

Robbins, T.W. and Sahakian, B.J. (1983) Behavioural effects of psychomotor stimulant drugs: clinical and neurophysiological implications. In: Creese, I. (ed.), *Stimulants: Neurochemical, Behavioural and Clinical Perspectives*. Raven Press, New York, pp. 301–337.

Robbins, T.W., Mittelman, G., O'Brien, J. and Winn, P. (1990) The neurophysiological significance of stereotypy induced by stimulant drugs. In: Cooper, S.J. and Dourish, C.T. (eds), *Neurobiology of Stereotyped Behaviour*. Clarendon Press, Oxford, pp. 25–63.

Robinson, T.E. (1988) Stimulant drugs and stress: factors influencing individual differences in the susceptibility to sensitization. In: Kalivas, P.W. and Barnes, C. (eds), *Sensitization of the Nervous System*. Telford Press, Caldwell, New Jersey, pp. 145–173.

Robinson, T.E. and Becker, J.B. (1986) Enduring changes in brain and behaviour produced by chronic amphetamine administration: a review and evaluation of animal models of amphetamine psychosis. *Brain Research* 11, 157–198.

Sahakian, B.J. and Robbins, T.W. (1975) The effects of test environment and rearing condition on amphetamine-induced stereotypy in the guinea pig. *Psychopharmacology* 45, 273–281.

Sharp, T., Zetterstrom, T., Herreea-Marschitz, M., Ljungberg, T. and Ungerstedt, U. (1986) Intracerebral dialysis – a technique for studying dopamine release in the rat brain in relation with behaviour. In: Joseph, M.H., Fillenz, M., MacDonald, I.A. and Marsden, C.A. (eds), *Monitoring Neurotransmitter Release During Behaviour*. Ellis Horwood, Chichester, pp. 94–104.

Simon, H. and Le Moal, M. (1984) Mesencephalic dopaminergic neurons: functional role. In: Usdin, E., Carlsson, A. and Engel, J. (eds), *Catecholamines; Part B: Neuropharmacology and Central Nervous System – Theoretical Aspects*. Alan R. Liss, New York, pp. 297–307.

Stone, E.A. (1990) Noradrenergic receptor mechanisms in stress adaptation. In: Puglisi-Allegra, S. and Oliverio, A. (eds), *Psychobiology of Stress*. Kluwer Academic Press, Dordrecht, pp. 151–160.

Tassin, J.P., Stinus, L., Simon, H., Blanc, G., Thierry, A.M., Le Moal, M., Cardo, B. and Glowinsky, J. (1978) Relationship between the locomotor hyperactivity induced by A10 lesions and the destruction of fronto-cortical DA innervation in the rat. *Brain Research* 141, 267–281.

Terlouw, E.M.C., Lawrence, A.B. and Illius, A.W. (1992) Relationship between amphetamine and environmentally induced stereotypies in pigs. *Pharmacology, Biochemistry and Behaviour* 43, 329–340.

Ungerstedt, U. (1984) Measurement of neurotransmitter release by intracranial dialysis. In: Marsden, C.A. (ed.), *Measurement of Neurotransmitter Release 'In Vivo'*. John Wiley, Chichester, pp. 81–105.

Valenstein, E.S. (1976) Stereotyped behaviour and stress. In: Serban, G. (ed.), *Psychopathology of Human Adaptation*. Plenum Press, New York, pp. 113–124.

Vezina, P. and Stewart, J. (1989) The effect of dopamine blockade on the development of sensitization to the locomotor activating effects of amphetamine and morphine. *Brain Research*, 499, 108–120.

Weiss, J.M., Goodman, P.A., Losito, B.G., Corrigan, S., Charry, J.M. and Bailey, W.H. (1981) Behavioural depression produced by an uncontrollable stressor: relationship to norepinephrine, dopamine, and serotonin levels in various regions of rat brain. *Brain Research Review* 3, 167–205.

Whishaw, I.Q. and Mittelman, G. (1991) Hippocampal modulation of nucleus accumbens: behavioural evidence from amphetamine-induced activity profile. *Behavioural and Neural Biology* 55, 289–306.

Wiepkema, P.R. (1990) Stress: Ethological implications. In: Puglisi-Allegra, S. and Oliverio, A. (eds), *Psychobiology of Stress*. Kluwer Academic Press, Dordrecht, pp. 1–14.

Wise, R.A. (1980) The dopamine synapse and the notion of pleasure centre in the brain. *Trends in Neurosciences* 3, 91–95.

Yoshikawa, T., Shibuya, H., Kaneno, S. and Toru, M. (1991) Blockade of behavioural sensitization to methamphetamine by lesion of hippocampo-accumbal pathway. *Life Sciences* 48, 1325–1332.

Functional Consequences of Behavioural Stereotypy

ROBERT DANTZER[1] AND GUY MITTLEMAN[2]
[1]*INRA-IN SERM U176, Bordeaux Cedex, France:* [2]*Memphis State University, Memphis, USA.*

Editors' Introductory Notes:
In this chapter Dantzer and Mittleman consider in detail the function of behavioural stereotypy, a subject already raised several times in this book and one that has excited considerable debate over the years. In terms of their research interests, Dantzer and Mittleman occupy something of the middle ground between the pure biomedical approach presented in Chapter 6 and farm animal studies (e.g. Chapter 5).

The central issue Dantzer and Mittleman address is the hypothesis that stereotypies are more than a simple response to stress; indeed that stereotypies allow animals to cope with aversive circumstances over which they have little control. In their review they draw heavily on evidence from studies of schedule-induced behaviour (e.g. schedule-induced polydipsia (SIP)) and drug-induced stereotypies. We have retained a certain amount of overlap with previous chapters (e.g. Chapter 6), as the approaches adopted by the different authors give interestingly alternative perspectives on the same data.

Initial studies of SIP supported the coping hypothesis as performance of the behaviour was found to be correlated with reduced hypothalamic–adrenal–pituitary (HPA) activity. However, the picture has since been complicated by studies that have failed to clearly replicate these original findings. In summarizing the present situation, Dantzer and Mittleman suggest that glucocorticoids may play a permissive role in the early stages of development of the behaviour. Furthermore, certain results suggest that, although SIP may not develop as a coping response, it may become one (at least in terms of HPA activity) after high and sustained levels of behaviour are reached. There is also recent evidence that amphetamine-induced stereotypy can act to reduce HPA activity. However, as Dantzer and Mittleman point out, there are other viable explanations for the apparent stress-reducing properties of SIP. Furthermore, studies of other schedule-induced behaviour have generally failed to provide consistent support for the coping hypothesis.

On the surface this chapter appears to provide little support for the coping hypothesis. However, Dantzer and Mittleman conclude their review optimistically by suggesting that the picture that is emerging is simply of greater complexity than has previously been acknowledged. They suggest that the most promising approach to a fuller understanding of this complexity will be through the study of individual coping strategies, thus leading to more in-depth research and a coordination between different specialities.

Introduction

Stereotypies are repetitive actions that are fixed in form and orientation and serve no obvious function. By proposing this definition, the group of experts that met in Brussels at the initiative of the Commission of the European Communities a few years ago (CEC, 1983) intended to help scientists, technicians and laymen recognize what represents the most common and severe abnormal behaviour observed in intensively housed farm animals.

According to this definition, behavioural stereotypy serves no purpose. If this is indeed correct, it makes little sense to search for functional consequences of these activities beyond those directly resulting from the intense body or mouth movements that form the basis of stereotyped behaviour. In contrast to this opinion, the major purpose of this chapter is to suggest that not all occurrences of stereotyped behaviour are purposeless, and that in some situations stereotypy may have functional consequences for an animal. Furthermore, the study of these consequences may be of some help in understanding the nature of behavioural stereotypy.

Conceptual Issues

It is difficult to adhere to a strict definition of stereotyped behaviour as purposeless after considering its eliciting conditions or its significance in the time budget of animals. Stereotypy usually occurs under conditions of stress, conflict or frustration and is typically expressed at high rates, and for long periods of time. Based on these somewhat superficial considerations, it is tempting to attribute to behavioural stereotypy a compensatory role in an animal's economy (Dantzer, 1986). This conclusion is based on the reasoning that individuals placed in stressful situations are unlikely to remain passive. They will actively try to cope with the situation by engaging in activities enabling them either to change the nature of the eliciting conditions, or to withdraw from them. When this cannot be achieved directly, there are other means available to deal with the situation, such as engaging in displacement activities or redirecting the frustrated motivation towards another object. For those animals which have sufficiently elaborate mental representations, it may also be possible to reinterpret the situation in a different way, perhaps through denial or blunting, so that the eliciting conditions are no longer perceived as threatening. These various ways of responding to external challenges represent what psychologists typically refer to as coping strategies. Problem-focused coping refers to actions initiated to modify the eliciting situation, while emotion-focused coping applies to those intra-psychic processes, also known as defence mechanisms, which result in a different interpretation of the situation (Dantzer, 1989).

Within this context, stereotyped behaviour may be considered as serving a coping function, if it reduces the aversiveness of the eliciting conditions or perhaps alters levels of sensory stimulation to compensate for either a lack or abundance of environmental stimuli. In either case, the ultimate aim of the stereotyping animal could be conceived of as modulating its level of arousal to keep it at an optimum level. As used here, the construct of arousal refers to a non-specific state of neural activity which fluctuates according to time of day and variations in underlying behavioural and mental activities and which could be considered aversive when too high or too low.

Although the coping hypothesis, in this general form, is quite appealing, it is not easy to test, especially in situations where stressed individuals cannot directly control or alter their environment. It is necessary to question subjects either by asking them whether they feel better when they use a specific coping strategy or by measuring some internal variable indicative of their arousal level. In human beings, psychologists use questionnaires and interviews whereas physiologists prefer to use more objective measures such as plasma levels of stress hormones. In animals, research on coping is necessarily limited to physiological approaches. The logic behind this approach is that if a stressor elicits a physiological response 'a' and if, by engaging in a specific action, the observed individual displays a lower physiological response 'b', then, by definition, the action initiated by the subject is considered a coping response (Levine et al., 1978). Although this may appear a sound assumption, a potential problem is that the behaviour the coping individual engages in can also affect 'a' so that 'b' is actually the result of an interaction between this effect and the consequence of coping. This is particularly the case when metabolic and hormonal indicators are used since changes in the level of physical activity will frequently impact upon underlying metabolic processes (e.g. Leshner, 1978).

There are alternative ways, however, of stating that a given behaviour can serve a coping function. For example, it can be assumed that there are systems in the brain which, when stimulated, result in mental states opposing the distress or tension the subject is experiencing. These systems may be stimulated either endogenously, by positive thinking or engaging in certain activities (e.g. pleasant physical activities), or exogenously, by consuming drugs which have euphoric or hallucinogenic properties. In both cases, the physiological approach described previously is no longer applicable since the stimulation to which the subject is exposing himself can, itself, affect his internal milieu, perhaps to an even greater extent than the stressful situation itself. For example, it is very difficult to assess the validity of the hypothesis that alcohol consumption helps reduce stress by recourse to physiological variables. Alcohol itself induces increases in plasma levels of stress hormones comparable to those observed in stressed individuals (Dantzer and Ollat, 1991). In the same manner, within the context of classical stress theory, substituting joy for sorrow may not significantly alter physiological indicators of stress since both emotions are claimed to be accompanied by non-specific hormonal changes typical of the response to stress. On the basis of these different conceptual issues, it is clear that interpretation of the functional consequences of stereotypies is not as simple and straightforward as it would appear *a priori*.

Methodological Issues

If behavioural stereotypy functions as a coping response which reduces the aversiveness of the situation in which animals are placed, it should be accompanied by decreased levels of arousal in comparison to those exhibited by animals placed in the same situation that do not display similar behaviours. The way in which this prediction is tested depends very much on the possibility of comparing an appropriate number of animals in each group, with sufficient homogeneity between individuals and environmental conditions. Another important factor is the choice of a reliable and sensitive indicator of arousal.

In zoo animals, single case studies predominate and appropriate controls are difficult to establish. Longitudinal studies are theoretically possible although problems resulting from handling and restraining animals for repeated blood sampling act as strong deterrents for any type of physiological research. In farm animals, the incidence of stereotypies in individuals kept in confinement is usually high enough to enable statistical studies but heterogeneity in the amount and quality of stereotypy developed by different individuals is often too high to permit well-controlled studies.

For these reasons, experimental analogues of stereotyped behaviour are often studied in place of naturally occurring stereotypy. Different approaches have been used, ranging from investigation of focused stereotyped interactions with specific objects in the immediate environment (e.g. repetitive chain-chewing in tethered sows) to inducing stereotyped behaviour by pharmacological means.

Behavioural Stereotypy and Stress

A basic tenet of the coping hypothesis is that behavioural stereotypy occurs as a response to stress. Consequently it is logical first to determine the generality and validity of this assumption. There is substantial evidence that certain stressful environmental conditions, such as separate housing or rearing in isolation, produce stereotyped behaviour in zoo and laboratory animals (Meyer-Holzapfel, 1968; Ridley and Baker, 1982; Fentress, 1983). The adaptive significance of this behaviour is not clear. Fitzgerald (1967) suggested that stereotypies functioned to increase sensory input during periods of relative sensory deprivation. Such a hypothesis is not dissimilar to stereotypy as a coping response if it can be assumed that stress is accompanied by reduced, rather than heightened sensory stimulation. Ridley and Baker (1982), although conceding that behavioural stereotypy produced by early social isolation might serve to increase sensory stimulation, maintained that it was unlikely to serve any particular function and instead resulted from pathological changes in the brain. They showed that in monkeys stereotyped behaviour elicited by amphetamine was more similar to stereotypy induced by early isolation rather than by single housing. As the monkeys in this experiment were exposed to both conditions in this study (early social isolation and separate caging), these results show that different stressors produced different patterns of stereotyped behaviour in the same animal. This raises doubts about the generality of the assumption that the different eliciting conditions could all be considered as stressful. Alternatively, it may be reasonable to assume that these different conditions could simply result in different amounts of stress, which could, in turn, account for the observed morphological differences in behavioural stereotypy.

In rats, the similarities between the behaviour elicited by a variety of stimuli such as electrical stimulation of the lateral hypothalamus (ESLH), tail-pinch, intermittent schedules of reinforcement and social isolation have suggested to many investigators that all these stimuli are stressful and that the elicited behaviours are stereotyped. In response to all these situations rats display increases in locomotor activity and in many cases develop significant repetitive oral behaviours such as eating, licking and biting. These oral and locomotor activities at least superficially resemble the repetitive stereotyped locomotor and oral behaviours elicited by various doses of amphetamine. These eliciting conditions also cause biochemical and endocrinological changes that are similar to those observed following traditional stressors (e.g. Antelman and Caggiula, 1977). Thus, for example, tail-pinch, like classical stressors such as footshock and immobilization, causes increases in plasma corticosterone levels and noradrenaline turnover, and like other stressors, when repeatedly administered, can 'sensitize' an animal to a subsequent injection of amphetamine (Antelman and Chiodo, 1983, 1984).

Although some stressful situations may result in stereotyped behaviour it should be noted that other stressors, including cold, immobilization, and exposure to inescapable electric shocks, do not appear to elicit stereotypy, thus questioning the generality of the hypothesis that stress increases stereotypy. As suggested by Robbins et al. (1990), this difference may be related to differences in the stimulus properties of different stressors. Thus, one common feature of stressors that elicit stereotypy is that they often lack explicit environmental cues. In comparison to electric shock or changes in temperature which involve quite specific stimulation in addition to the stressful state, treatments such as isolation rearing, electrical stimulation of the lateral hypothalamus and tail-pinch, all lack strong exteroceptive properties. In this contest stereotypy may be a common feature of conditions involving high levels of stress or arousal, but with no obvious external cue as to the cause of stress.

Adjunctive Behaviour and Coping

The question of whether stereotyped behaviour may fulfil a coping function has been addressed in a number of different paradigms. An often used technique for evoking highly repetitive and stereotyped behaviour involves exposing food-restricted animals to an intermittent schedule of reinforcement and providing them, at the same time, with something to do in the intervals between food rewards. The best known example of this is the schedule-induced polydipsia (SIP) paradigm. In a now classic experiment (Falk, 1961), food-deprived rats were trained to respond on a 1 minute variable-interval (VI-1 min) schedule in which a lever press produced a small food pellet on average once a minute. After the rats had eaten each pellet they immediately began drinking water. Although post-prandial drinking is not remarkable, the quantity ingested was significantly above normal levels. For example, during a 3 hour test session, rats consumed more than three times their usual daily water intake. Falk called this phenomenon 'schedule-induced polydipsia' and subsequently reported that it could be produced by a variety of different reinforcement schedules including regular delivery of a food pellet at 60 second intervals without the requirement that the animal make an operant response (Falk, 1961, 1966).

Falk (1966) suggested that SIP should be classified as an 'adjunctive behaviour' because it: (i) occurs as an adjunct to a reinforcement schedule; (ii) is not directly involved in or maintained by the reinforcement contingency; but (iii) has reinforcing properties. It has subsequently been demonstrated in rats that many behaviours besides drinking can be elicited by the intermittent delivery of a food reward, including attack (Azrin, 1961; Gentry, 1968; Hutchinson et al., 1968), pica-eating (Freed and Hymowitz, 1969), air-licking (Taylor and Lester, 1969; Mendelson and Chillag, 1970), wheel-running (Levitsky and Collier, 1968; Wallace et al., 1983), escape (Azrin, 1961;

Thompson, 1964) and increased activity (Killeen, 1975). In humans, adjunctive behaviour in the form of locomotor activity, polydipsia, cigarette smoking, and eating (Kachanoff *et al.*, 1973; Wallace and Singer, 1976; Cherek and Brauchi, 1981) has been reported. Falk (1971) has argued that all of these different behaviours belong to the class of adjunctive behaviour because they share many common properties. He noted the following commonalities: (i) the rate of occurrence of these behaviours is an inverted U-function of the inter-reinforcement interval; (ii) the rate of occurrence of these responses decreases as an animal's food deprivation level decreases; (iii) the behaviour is persistent and excessive; (iv) the time immediately after reinforcement is the temporal focus of the behaviour; (v) there is no response-reinforcement contingency; (vi) animals will learn an operant response in order to engage in these behaviours; (vii) both the frequency of occurrence of the behaviour and the form that it takes are affected by environmental stimuli.

Perhaps the best known hypothesis of SIP, as well as other adjunctive behaviours, is that they are laboratory analogues of the displacement behaviour observed by ethologists (Falk, 1971, 1977). Both adjunctive and displacement behaviours occur in situations where a strongly motivated appetitive or consummatory behaviour is interrupted or prevented. Both types of behaviour are characterized by their exaggerated or excessive nature, and their adaptive value with respect to the impeded behaviour is questionable. It has been suggested that adjunctive behaviours are adaptive as they permit animals to remain engaged in a situation that is favourable to their survival, but also contains an aversive component (Falk, 1971, 1977). Accordingly, a hungry rat that receives food on an intermittent schedule has its hunger partially appeased, but its feeding behaviour is also repeatedly thwarted. In this situation, drinking develops from the conflict between two opposing tendencies; feeding behaviour (a response to the rewarding components) and escape behaviour (a response to the aversive components). According to this argument, drinking is adaptive because it serves a stabilizing or buffering function.

Unfortunately, this is a difficult hypothesis to test. However, the essential element of this view, that SIP develops from a conflict between two incompatible response tendencies, led to the development of a more empirical hypothesis. As suggested by Killeen *et al.* (1978), the activation or arousal that accompanies the delivery of a scheduled food pellet may outlast its consumption, and potentiate alternative activities (i.e. drinking) evoked by available environmental stimuli. These alternative activities are performed vigorously until interrupted by competing behaviour emitted in anticipation of the next food pellet. In support of this view it was demonstrated that the presentation of a single food pellet generated a small amount of behavioural activation, as indicated by increases in activity counts, which then decayed exponentially over time. When successive food pellets were delivered,

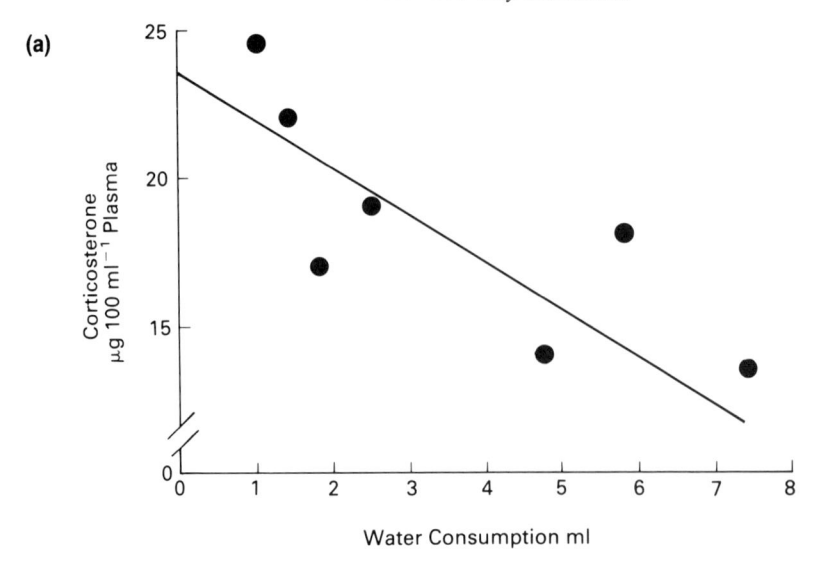

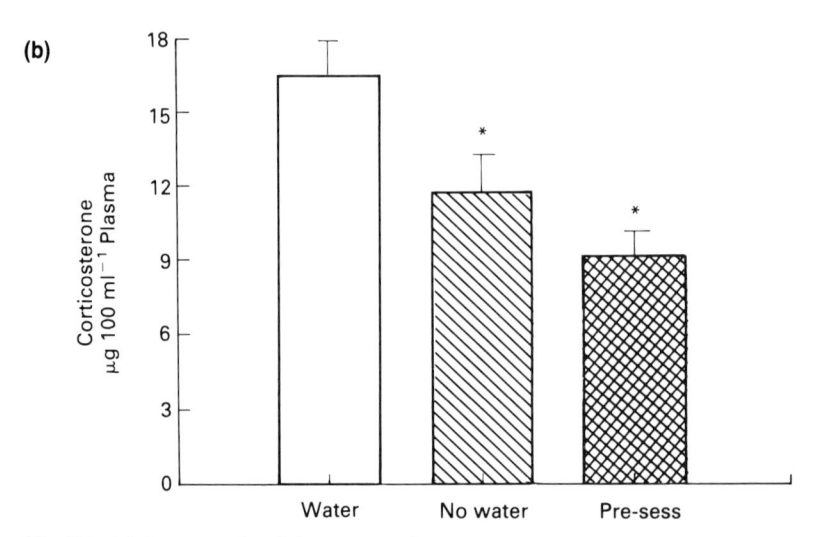

Fig. 7.1. (a) A scatter plot of the amount of water consumed during a 30-minute SIP test session as a function of plasma corticosterone levels. Rats in this figure had received a total of ten test sessions (one test session per day). Corticosterone levels were determined immediately following the end of the final test session. (b) Plasma corticosterone levels of animals exposed to a fixed-time schedule of food reinforcement with water available from a drinking tube (Water) or with drinking prevented by the removal of the water tube (No water). Plasma corticosterone levels were determined immediately following the 30-minute test. All groups had received a total of ten, SIP test sessions. Plasma corticosterone levels were determined in the Water and No water groups immediately following the final test, while corticosterone levels in the pre-session (Pre-sess) group were determined prior to the

behavioural activation accumulated and was then maintained at an asymptotic level. However, the potential activation or activity reducing effects of drinking were not investigated.

Using a different measure, Brett and Levine (1979) demonstrated the adaptive role of drinking when they reported that exposure to scheduled food delivery elevated plasma corticosterone levels, but that drinking suppressed this elevation. If animals were not allowed to drink during SIP training sessions, corticosterone levels returned to the pre-test session baseline. The low levels of plasma corticosterone associated with SIP could not be attributed to simple haemodilution because significant reductions in corticosterone were observed even when the amount of water consumed was relatively small (Brett and Levine, 1981). These corticosterone-reducing effects of SIP have been replicated (Tazi et al., 1986) and it has further been shown that if SIP is prevented by removal of the drinking tube, then plasma corticosterone levels are significantly elevated in comparison to those seen when SIP is permitted (Tazi et al., 1986). Enhanced plasma corticosterone levels are also found in those rats which do not drink, despite having access to water (Dantzer et al., 1988a). As it is widely assumed that circulating concentrations of pituitary–adrenal hormones such as corticosterone are reliable indicators of arousal (Hennessy and Levine, 1979), these results are consistent with schedule-induced drinking serving a coping function by reducing stress.

However, the view that SIP functions as a coping response that reduces corticosterone levels has become controversial as results inconsistent with this hypothesis have also been reported. Rather than decreasing corticosterone levels, it has been demonstrated that SIP can sometimes significantly increase plasma corticosterone compared to baseline levels (Wallace et al., 1983, Table 4, p. 134), or compared to levels exhibited by animals exposed only to the schedule of food presentation but not permitted to drink (Fig. 7.1). Furthermore, exposure to scheduled food delivery with water unavailable does not increase plasma corticosterone above basal levels (Tazi et al., 1986, Figure 3, p. 252). Additionally, as shown in Fig. 7.1, individual animals which fail to develop SIP have high plasma corticosterone levels, suggesting that rather than SIP modulating corticosterone, plasma corticosterone is involved in the modulation of SIP (Mittleman et al., 1988).

The relationship between pituitary–adrenal activity and SIP has been

---

start of the final test session. *$P<0.05$, Newman–Keuls comparisons between the Water group and either the No water or Pre-sess groups. It should be noted that the results depicted in the top and bottom of this figure are problematic for the coping hypothesis of SIP. Specifically, the top figure indicates that high levels of corticosterone are related to low levels of SIP which may be suggestive of a modulatory role of corticosterone in the development of SIP. The lower figure shows that prevention of SIP by removal of the opportunity to drink can reduce, rather than augment, plasma corticosterone levels. (From Mittleman et al., 1988.)

further investigated by eliminating adrenal glucocorticoids with adrenalectomy, again with conflicting results. Thus, Devenport (1978) found that adrenalectomy significantly facilitated the onset of SIP while adrenal demedullation was without effect, suggesting that greatly reduced levels of corticosterone are associated with enhanced acquisition of this behaviour. In contrast, Levine and Levine (1989) have reported that adrenalectomy significantly retards the acquisition of SIP when measured as fluid consumption as well as when using measurements similar to those of Devenport (1978). Additionally, it has been reported that adrenal demedullation, but not adrenalectomy, suppresses the acquisition of SIP (Wright and Kelso, 1981).

These conflicting findings highlight at least one question that is crucial for determining if SIP as well as other adjunctive behaviours serve a coping function. As investigations of SIP and measures of pituitary–adrenal function have provided only inconsistent support for the coping hypothesis, it is reasonable to ask if there is a consistent relationship between SIP and pituitary–adrenal activity. The consistency of the relationship between SIP and plasma corticosterone has been recently investigated (Mittleman et al., 1992). A series of experiments compared the effects of increased or decreased plasma corticosterone levels on the acquisition of performance of SIP. The results indicated that the acquisition of SIP could be decreased by adrenalectomy, blockade of corticosterone synthesis, or administration of corticosterone. Performance of established SIP was also decreased by adrenalectomy. The effects of corticosterone administration on established SIP depended on the level of performance. High levels of drinking were enhanced by a high dose of corticosterone while low rates of drinking were increased by a low dose of corticosterone. These results were interpreted as being consistent with plasma corticosterone playing a permissive role during the acquisition of SIP. However, once drinking reached high and consistent levels, SIP may well become an effective coping response in that under some situations, high levels of plasma corticosterone at the start of a test session could lead to increased SIP, while low levels lead to decreased SIP. Stated differently, the results reviewed above are broadly consistent with the notion that although SIP does not appear to develop as a coping response it may well become one after high levels of performance are reached. Successful demonstrations of the plasma-corticosterone-reducing properties of SIP have only been obtained in rats that are either extensively trained and/or drinking large volumes of water (Brett and Levine, 1979; Tazi et al., 1986; Dantzer et al., 1988a). From the results of Mittleman et al. (1992), it would be expected that these animals are the most likely candidates to show such a relationship. Thus it may not be surprising that rats drinking smaller amounts during SIP do not show similar reductions in plasma corticosterone (Wallace et al., 1983; Mittleman et al., 1988).

The consistency of the relationship between adjunctive behaviour and pituitary–adrenal activity has also been investigated using other paradigms. Pigs which are exposed to an intermittent delivery of food with a chain in front of them develop chain-manipulation in the interval between food deliveries and display lower plasma cortisol levels than in the absence of the chain (Dantzer and Mormède, 1981). This significant drop in plasma cortisol occurs only when the animals are fed intermittently and not when they receive all of their food at the beginning of a test session (Dantzer et al., 1986b). This decrease in plasma cortisol levels observed in pigs engaged in chain-manipulation occurs only in comparison to pre-session levels (Dantzer and Mormède, 1981; Dantzer et al., 1986b). Although this demonstration is suggestive of a stress-reducing role for chain-manipulation it should be noted that chain-manipulation shares only some properties with adjunctive behaviour. Specifically, the intensity of chain-chewing is similar in animals given all their food at the beginning of the experimental session to those receiving food intermittently (Dantzer et al., 1986b; see also Terlouw et al., 1991)

The potential coping effects of schedule-induced wheel-running have also been investigated. Rats given food on an intermittent schedule with access to a running wheel exhibited increases in plasma corticosterone levels in direct relation to the amount of running performed (Tazi et al., 1986).

Other physiological indices of stress or activation have been investigated with similarly inconsistent results. Dantzer et al. (1988a) catheterized rats and then periodically withdrew blood samples during test sessions when SIP was being acquired. It was found that although plasma corticosterone levels declined within the sessions in drinking rats, plasma concentrations of catecholamines and prolactin failed to show similar reductions and instead remained constant during the same sampling period. Thus, different indices of stress showed different results.

If we assume that adjunctive behaviours do not fulfil a coping or stress-reducing function, it is logical to investigate what other mechanisms can account for the reduction in plasma corticosterone levels. Useful information on this question has come from studies on the effects of feeding schedules on pituitary–adrenal activity (Heybach and Vernikos-Danellis, 1979a, b; Honma et al., 1986). In rats and perhaps in other species as well, restriction of feeding to a single daily meal alters the circadian rhythmicity of pituitary–adrenal activity. There is a shift of the normal afternoon peak in plasma corticosterone concentrations to the time just before the daily feeding session. Plasma levels of corticosterone as well as adrenocorticotrophin hormone (ACTH) fall rapidly upon presentation of food and water. This rapid fall can also occur in response to the cues that have been previously associated with the daily presentation of food and water. However, sustained decreases in pituitary–adrenal activity are dependent on actual food or water consumption. Consummatory activities appear to have potent suppressive influences on the pituitary–adrenal axis, not only directly at the level of the hypothalamus and the pituitary, but also indirectly via some unknown mechanism controlling the clearance of circulating corticosteroids. Two sets of data support this last possibility. First, plasma corticosterone decreases significantly during SIP in adrenalectomized rats implanted with a corticosterone pellet (Levine and Levine, 1989); second, the decline in plasma corticosterone concentrations that results from drinking occurs too rapidly (2–3 minutes) to be accounted for by pituitary or hypothalamic influences (Wilkinson et al., 1982). Taken together, these results suggest that the decline in plasma corticosterone levels may be attributed, at least in part, to metabolic mechanisms rather than to psychological or coping processes.

From the literature reviewed above it is apparent that there is only very weak support for the hypothesis that adjunctive behaviour serves a coping function and that other viable explanations of the relationship between physiological indices and SIP may well exist. Experiments on SIP have yielded inconsistent, although potentially explainable, results. Similarly conflicting results have also been obtained when other forms of adjunctive behaviour or other physiological indices of stress have been investigated. Thus, in essential agreement with the initial remarks of Falk (1971) concerning schedule-induced polydipsia, there still appears to be no single acceptable behavioural account of this seemingly bizarre phenomenon. It remains possible, however, that SIP may come to serve a coping function once this behaviour has reached high and consistent levels.

Stimulant-induced Stereotypy and Coping

Stereotyped behaviour can be reliably induced by pharmacological treatment with dopaminergic agonists such as amphetamine or apomorphine (cf. Chapter 6). These agents are injected into animals and stereotypy is assessed by visual inspection and/or activity recordings. This technique is acute in the sense that bouts of stereotyped behaviour occur during the time course of activity of the drug injected, lasting only for a few hours. The question of whether stereotypy induced by dopamine agonists may fulfil a coping function rests at least partly on the assumption that these drugs are stressors and that stereotypy represents a coping response to their stressful effects.

That amphetamine is a stressor has been well argued by Antelman and Chiodo (1983) who compared the physiological, neuroendocrine and neuropharmacological effects of psychostimulant drugs and conventional stressors. They postulated that for amphetamine to be considered a stressor, it should act in an additive and substitutable way in combination with other stressors. Specifically they suggested that: (i) the presence of stress should increase the animal's sensitivity to amphetamine in terms of stereotypic behaviour and therefore move the dose–response curve for stereotypy to the left; (ii) sensitization to the effects of prior amphetamine injection could be replaced by previous experience with other stressors. In support of these points they demonstrated that previous experience with tail-pinch, footshock, or food deprivation can enhance subsequent amphetamine-induced stereotyped behaviour. While there is evidence that the stressors Antelman and Chiodo (1983) describe can alter the behavioural response to amphetamine and other dopamine agonists, certain limitations or extensions of this proposition should be noted. For example, MacLennan and Maier (1983) have shown that exposing rats to electric footshocks enhanced the subsequent response to amphetamine at a dose of 4 mg kg^{-1}, only if the shocks were uncontrollable (precluding 'coping responses'). In this experiment, rats that received the same number of shocks as control animals, but were allowed to escape from them, showed levels of stereotypy similar to those of unshocked rats. In comparison, the group that had received uncontrollable shocks showed significantly higher levels of stereotypy than the other groups. The ability to control a stressor has also been investigated in a similar experiment using corticosterone as a neuroendocrine index of stress (Swenson and Vogel, 1983). As shown in Fig. 7.2, rats that had received a series of uncontrollable footshocks (Non-coping group) exhibited a significantly prolonged increase in plasma corticosterone in comparison to a yoked group given the opportunity to escape (Coping group). Taken together, these studies suggest the opportunity to control a stressor has a large influence in determining the stressfulness of a situation and its subsequent impact on amphetamine-induced stereotypy.

More generally, it should be noted that the proposition that amphetamine

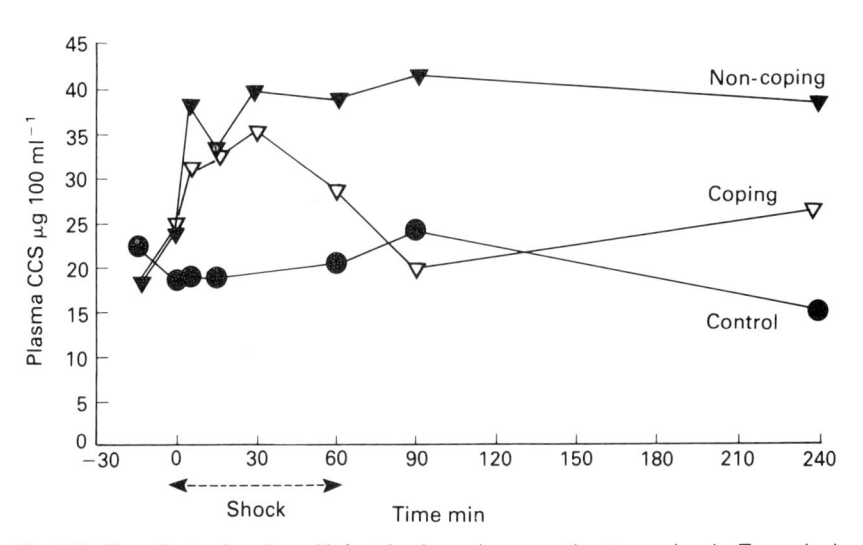

Fig. 7.2. The effects of coping with footshock on plasma corticosterone levels. Two yoked groups of rats received 60 minutes of unsignalled intermittent footshock (1 mA, total duration approximately 500 seconds) that was escapable (Coping group) or inescapable (Non-coping group). The intermittent footshock occurred during the first 60 minutes of the test session. In non-coping rats, the increase in plasma corticosterone levels due to shock was significantly prolonged in comparison to animals permitted to cope with shock as well as compared to an unshocked control group. (From Swenson and Vogel, 1983.)

is a stressor is necessarily limited, due to the exclusive use of stereotypy rating scales to measure the effects of amphetamine as well as the lack of full dose-response curves. Thus, for example, it is well known that food deprivation can increase the locomotor effect of amphetamine (Campbell and Fibiger, 1971) and apomorphine (Sahakian and Robbins, 1975). However, from rating scale measures alone, it is very difficult to determine if this type of behavioural alteration is similar to the changes in stereotypy described by Antelman and Chiodo (1983). As the stereotypy induced by amphetamine or apomorphine is also differentially altered by exposure to stressors such as isolation rearing and food deprivation (Campbell and Fibiger, 1971; Sahakian and Robbins, 1975; Sahakian *et al.*, 1975), concerns such as this become more important. Additionally, as it is well known that different neurophysiological substrates underlie the main behavioural components of the response to amphetamine (Randrup and Munkvad, 1967; Creese and Iversen, 1975; Kelly *et al.*, 1975; Costall and Naylor, 1977), considerations such as these strongly suggest that the relationship between specific aspects of the amphetamine response and stress requires further explanation.

There is relatively little information concerning the question of whether pharmacologically induced behavioural stereotypy subserves a coping function. According to the coping hypothesis, supporting evidence could involve demonstrations that amphetamine-induced stereotypy leads to the reduction in the neuroendocrine concomitants of the 'stressful' effects of the drug, such as elevated plasma corticosterone. It is immediately apparent that this is a very difficult task. If stereotyped behaviour were blocked, for example, by pharmacological means, then the coping hypothesis would predict that plasma corticosterone levels would be elevated above normal for that dose of amphetamine.

A recent experiment has investigated the relationship between amphetamine-induced stereotypy and coping, again using plasma corticosterone levels as a neuroendocrine index of arousal (Jones *et al.*, 1989). Dopamine-depleting lesions of the caudate putamen were used to block amphetamine-induced stereotypy. Results of this experiment are shown in Fig. 7.3. These lesions significantly prolonged the increase in plasma corticosterone associated with 5 mg kg^{-1} d-amphetamine. In contrast, similar dopamine-depleting lesions of the nucleus accumbens, which blocked the locomotor, but not the stereotypic response to amphetamine, did not have this effect. These results clearly supported the hypothesis that amphetamine-induced stereotyped behaviour serves a coping function and specifically implicated the dopamine system that originates in the mesencephalon and projects to the striatum in the modulation of the corticosterone response. As shown in Figs 7.2 and 7.3, this prolongation of the increase in plasma corticosterone produced by amphetamine strikingly resembled the lengthened corticosterone response of rats submitted to a series of footshocks without the opportunity to control them (Swenson and Vogel, 1983).

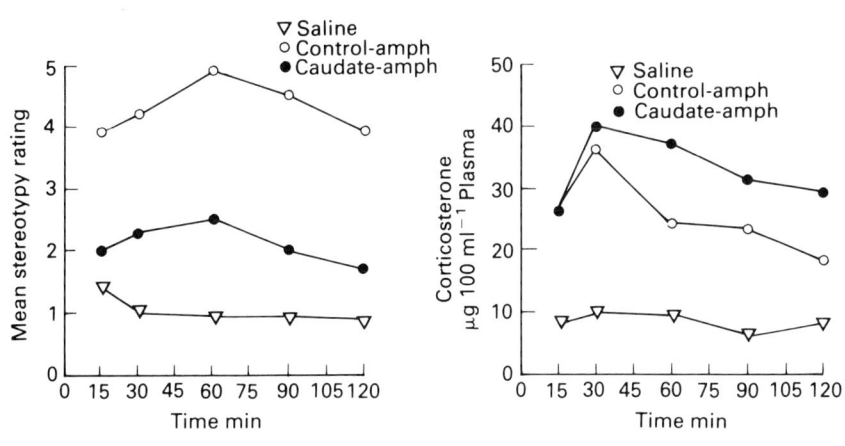

Fig. 7.3. The time course of behavioural stereotypy and plasma corticosterone levels induced by an injection of d-amphetamine (5 mg kg^{-1}) in rats with bilateral dopamine-depleting lesions of the caudate putamen (Caudate-amph) and unoperated animals (Control-amph). In addition, average levels of stereotypy and corticosterone in unoperated rats that received an injection of saline are shown (Saline). Stereotypy was rated on the following scale: 0 = inactive: 1 = intermittent locomotor activity; 2 = continuous locomotor activity; 3 = intermittent stereotypy; 4 = continuous stereotypy over a wide area; 5 = continuous stereotypy in a restricted area; 6 = pronounced stereotypy in a restricted area; 7 = intermittent stereotyped licking or biting directed at the walls or the floor; 8 = continuous stereotyped licking or biting. In comparison to the other groups, blockade of stereotyped behaviour in the caudate-putamen lesioned rats significantly prolonged the increase in plasma corticosterone. Note the similarity of the lengthened plasma corticosterone response in the caudate group to that shown in Fig. 7.2. (Adapted from Jones et al., 1989.)

Although providing support for the coping hypothesis, this experiment left open the possibility that the changes in plasma corticosterone were a primary result of the lesion and only secondarily due to the blockade of stereotypy, as both the behavioural and endocrine responses were dependent on dopamine depletion in the caudate putamen. A second experiment has been conducted in order to evaluate this possibility and to further investigate the possible function of amphetamine-induced stereotypy by adopting a complementary strategy to that used by Jones et al. (1989). Rather than blocking stereotypy and measuring the increment in plasma corticosterone, this experiment determined the effects on corticosterone, of enhancing stereotypy using repeated amphetamine administration. A series of five injections of d-amphetamine was used to enhance stereotyped behaviour in control animals as well as rats with bilateral dopamine-depleting lesions of the caudate putamen (Mittleman et al., 1991). As shown in Fig. 7.4, this regimen of amphetamine injections significantly increased stereotyped behaviour and also reduced the normal elevation in corticosterone produced by treatment

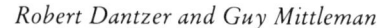

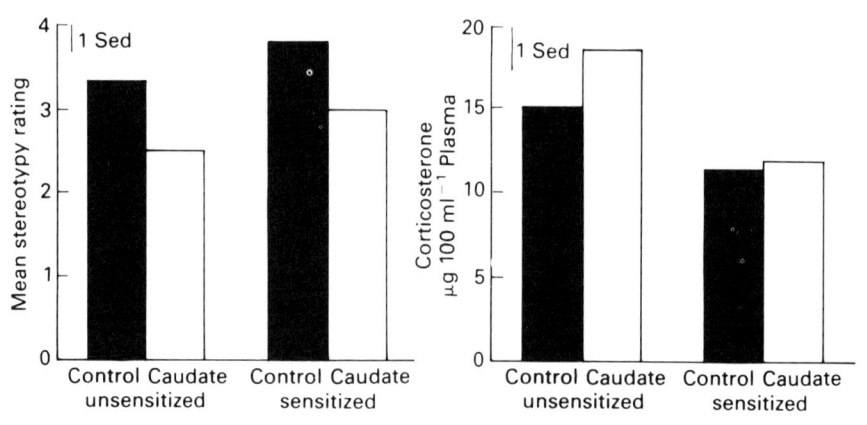

Fig. 7.4. Mean amphetamine-induced (dose = 5 mg kg^{-1}) stereotypy and plasma corticosterone levels collapsed across the 2 hour test interval in rats that had received either a series of five injections of d-amphetamine (sensitized) or saline (unsensitized). Both groups contained unoperated rats (Control) and animals with bilateral dopamine-depleting lesions of the caudate putamen (Caudate). The same stereotypy rating scale as in Fig. 7.3 was used. This regimen of amphetamine injections significantly increased stereotyped behaviour and also reduced the normal elevation in corticosterone produced by treatment with d-amphetamine. This effect was apparent in both control and lesioned animals. (From Mittleman *et al.*, 1991.)

with d-amphetamine. This effect was apparent in both control and lesioned animals. Taken together, the results of these experiments showed that reductions in stereotypy were associated with enhanced plasma corticosterone and that increases in stereotypy were related to decreases in corticosterone. Both sets of results provide strong support for the hypothesis that amphetamine-induced stereotyped behaviour functions to reduce stress or arousal, and additionally they suggest that this effect is largely independent of underlying dopaminergic mechanisms.

Physiological and Metabolic Aspects of Stereotypies

The physiological and metabolic concomitants of stereotypies can reflect both the metabolic cost of the corresponding physical activity and the coping function of these behavioural patterns. It is not always easy to dissociate these two components.

In well trained rats submitted to an intermittent schedule of food delivery, body temperature, monitored by implanted transmitters, increased by about 1°C in the course of 30-minute sessions. Although there was a slight tendency for drinkers to show less of an elevation than non-drinkers, this difference did not reach significance. The increase in body temperature was not a function of general activity since animals forced to run in a wheel did not show a higher hyperthermic response (Seguy, 1990).

In a systematic study of physiological correlates of SIP in rats submitted to an intermittent schedule of food distribution, Mittleman *et al.* (1990) observed that oxygen consumption was higher in drinkers than in non-drinkers. However, heart rate was lower in the drinkers which also tended to be less active than the non-drinkers. It would therefore appear that schedule-induced drinking is associated with inhibition of sympathetic nervous system activity, which should normally decrease energy mobilization. However, ingesting and then heating large quantities of room temperature water to body temperature before excreting it results in excessive energy expenditure which cannot be counterbalanced by the decreased sympathetic activity.

There has been little interest in the direct investigation of sympathetic nervous system activity in animals submitted to environmental conditions that favour the development of stereotyped behaviour, in spite of the fact that this is relatively easy to do. For example, tethering of adult male cynomolgus macaques had been shown to result in persistent elevations in heart rate compared to other stressful conditions. Administration of propranolol, a beta-adrenergic antagonist, induced an abrupt, sustained decrease in heart rate, indicating that the increase in heart rate associated with tethering was due to persistent stimulation of the sympathetic nervous system (Adams *et al.*, 1988).

The possibility that stereotypies induce a shift from sympathetic nervous system arousal to parasympathetic nervous system activation is intriguing and would be worth testing more systematically (cf. Chapter 3). It is supported by data from both infants and children. In infants, oral activities involved in sucking a pacifier activate vagal mechanisms, as deduced from the variation in the plasma pattern of gastrointestinal hormones (Uvnas-Moberg, 1987; cf. Chapter 5). In children moving from kindergarten to primary school, there is an increase in leg-swinging stereotypies during the time children have to remain seated at their desks and learn writing, reading and arithmetic. These stereotypies are accompanied by small but significant reductions in heart rate (Soussignan and Koch, 1985). Such a shift to parasympathetic activation may account for the lower incidence of abomasal damage in calves having developed tongue-playing stereotypies (Wiepkema *et al.*, 1987).

Stereotypies and Endogenous Opiates

Pharmacologically induced stereotypies can be attenuated by injection of naloxone, a specific antagonist of opiate receptors (Moon *et al.*, 1980). The same antagonist is able to block the acquisition of SIP but has no effect on its performance once fully developed (Riley and Wetherington, 1987). The expression of stereotypies in sows is also blocked by naloxone, at doses which minimally interfere with spontaneous behaviours such as exploration (Cronin et al., 1985, 1986).

In view of the abundance of data showing modulatory influences of endogenous opiates on dopaminergic neurons which are likely to be involved at some level in the development of stereotypies (cf. Chapter 6), it is quite surprising that the pig data on naloxone gave rise to such wild speculations about the rewarding or self-narcotizing properties of stereotypies in applied ethology (Cronin et al., 1985; Zanella et al., 1991). This interpretation emphasizes the coping value of stereotypies in terms of the mental state the stereotyping animal is able to achieve (euphoria or self-narcotization).

Actually, besides the observation that naloxone blocks developing stereotypies, there is little evidence that stereotypies increase endogenous opiates. Because of the complexity of distribution of endogenous opiate systems and the diversity of their molecular forms, this possibility has been investigated by using indirect measures such as pain sensitivity. It is well known that exposure to various stressors activates endogenous analgesic systems (Dantzer et al., 1986a), some of which are opiate-dependent in the sense that the stress-induced reduction in pain sensitivity is blocked by naloxone or other antagonists of opiate receptors. If stereotypies stimulate endogenous opiate systems, pain sensitivity should vary in the same direction as if animals were injected with an opiate agonist such as morphine i.e. stereotyping animals should display lowered sensitivity to pain as compared to non-stereotyping animals. Actually, the opposite changes were observed in rats having developed excessive drinking (Tazi et al., 1987) as well as in stereotyping sows (Rushen et al., 1990), in spite of the fact that, in both cases, exposure to the test conditions induced analgesia. In addition, the mere observation of a blocking effect of naloxone on stereotypies cannot be interpreted as the demonstration that this behaviour has rewarding properties. If it were the case, naloxone treatment should first increase the intensity and frequency of stereotypies since they would be experienced as less rewarding, subsequent to extinction taking place.

Stereotypies and Coping Strategies

There are two important issues to be considered in this section. The first one is that stereotypies do not develop with the same probability in different individuals exposed to the same eliciting conditions. The second one is that performance of stereotypies might interact with these individual factors to detrimentally influence brain functions.

The existence of individual risk factors predisposing to development of stereotypies is a relatively new concept. Until the mid-1980s, there was no mention in the literature of individual differences in response of laboratory animals or farm animals to pharmacological or environmental challenges. Cronin and Wiepkema (1984) showed for the first time that sows differed greatly in terms of the form, repetitiveness and intensity of their stereotypies, and that stereotypies of older sows were of a different nature to stereotypies of younger sows. At the same time, Mittleman and Valenstein (1984) initiated a series of systematic studies on the neurobiological basis of individual differences in schedule-induced activities. Rats that spontaneously display excessive drinking (SIP-positive animals) when exposed to an intermittent distribution of food and those that do not (SIP-negative animals) were compared both within SIP sessions and in other experimental situations. The main result of these studies was the observation of a consistent relationship between the predisposition to develop SIP and the propensity to respond by eating or drinking in response to electrical stimulation of the lateral hypothalamus (ESLH). As brain catecholamines, and especially dopamine, have been implicated in the regulation of oral activities elicited by exposure to a wide variety of mild stressors (cf. Chapter 6), they suggested that differences in predisposition to display SIP and ESLH-induced drinking might be related to individual differences in the responsiveness of forebrain dopaminergic systems. They went on to test this hypothesis by comparing the behavioural response of drinkers and non-drinkers to amphetamine, a dopaminergic stimulant, and their dopaminergic response to footshock stress. In both cases, SIP-positive rats showed a higher response, indicating that their central dopaminergic neurons were sensitized (Mittleman and Valenstein, 1984; Mittleman et al., 1986). Moreover, repeated injections of amphetamine were found to transform ESLH-negative rats into ESLH-positive animals (Mittleman and Valenstein, 1984). On the behavioural side, SIP-positive animals displayed more rapid acquisition of active avoidance in a shuttle-box and less freezing when confronted with an aggressive resident male in a defeat test than SIP-negative rats (Dantzer et al., 1988b).

These findings can be interpreted to suggest that the predisposition to develop SIP is another facet of a more general profile of behavioural and neurochemical reactivity to aversive situations. The behavioural characteristic of this profile could be a reduced ability to shift motor programs, or behavioural rigidity. SIP-positive animals may be individuals who easily develop routines and become more stereotyped in their way of responding to a given stimulus situation. This could reflect a greater predisposition to sensitization of the neural structures underlying the behavioural response.

Similar mechanisms are likely to account for the development of compulsive oral and locomotor stereotypies displayed by farm and zoo animals under conditions of conflict and frustration. These abnormal activities can appear as remnants of displacement activities or defensive reactions that are initially emitted to control the eliciting situation. In other situations, they have the appearance of fragmented elements of behavioural acts belonging to the repertoire of the thwarted motivation (Mason, 1991; Chapter 3). Normally, goal-directed behaviours have two components, an appetitive sequence enabling the organism to reach the goal object, and a consummatory phase, resulting in appropriation of the goal object. The appetitive phase is controlled by a positive feedback system enabling the organism to maintain the behaviour until the goal object is reached or higher priority activities inhibit the on-going behaviour. Consummatory activities provide a negative feedback to this system, the efficacy of which is dependent on the correct match between the actual result of behaviour and initial expectancies. The problem with stereotyped activities is that since they belong to the appetitive repertoire, they cannot be easily stopped in the absence of a suitable goal object. In addition, the underlying neural system is bombarded by intense oral and/or proprioceptive sensory feedback originating from performance of stereotypies, which contributes to activation. If the neuronal elements which compose the repeatedly activated neural pathways are prone to sensitization, all the conditions are met for a positive feedback in which the sensory factors that normally guide behaviour trigger a behavioural sequence which becomes self-organized independently of further environmental guidance or any particular motivational state (Dantzer, 1986).

This is nothing more than a restatement of principles of neural functioning already enunciated by Teitelbaum on the basis of his minute observations of animals recovering from brain lesions: 'If we assume that conflict can inactivate the higher-level neural controls characterizing the appetitive components of an instinctive behaviour chain, then the very same act may become less encephalized (more stimulus-bound, stereotyped, synergistic, and exaggerated in its intensity) as the behaviour is constantly repeated' (Teitelbaum, 1977, p. 23). In other terms, stereotypies would be both the external manifestation and the causal factor of functional brain alterations involving mainly the basal ganglia and their neurochemical modulation (Insel, 1988; Rapoport and Wise, 1988).

Conclusion

A cursory view of the present chapter would appear to give little support for the possibility that stereotypies have meaningful functional consequences. If it is possible to measure changes in a number of hormonal and physiological variables during or after stereotypic bouts, the knowledge which has been gained from such investigations seems to be quite limited.

This pessimistic conclusion is not warranted for a number of reasons. First, many interesting findings have emerged, from the demonstration of rapid changes in plasma corticosterone concentrations which are independent from central influences to the possibility of shifts in autonomic nervous activity during certain motor movements involving sensory feedback from either the mouth or the joints. Second, the accumulated data clearly demonstrate that unidimensional energetic constructs such as arousal, activation and motivation, are of limited use in explaining stereotypies. Complexity has to be accepted as such rather than being masked by simple constructs. We have now entered a new generation of studies on stereotypies in which the emphasis is on inter-individual differences in the propensity to develop such activities in a given situation. The objective is to compare stereotyping animals to non-stereotyping ones, not only in terms of their possible differences during development or maintenance of stereotypies, but also beforehand, in order to determine what are the risk factors that predispose to stereotypies. Such research cannot be carried out only from the surface and with the different perspectives isolated from each other because of the specialized discipline they belong to. It requires in-depth studies and integration therefore representing a challenge for future studies.

References

Adams, M.R., Kaplan, J.R., Manuck, S.B., Uberseder, B. and Larkin, K.T. (1988) Persistent sympathetic nervous system arousal associated with tethering in cynomolgus macaques. *Laboratory Animal Science* 38, 279–281.

Antelman, S.M. and Caggiula, A.R. (1977) Tales of stress-related behaviour: a neuropharmacological approach. In: Hannin, I. and Usdin, E. (eds), *Animal Models in Psychiatry and Neurology*. Pergamon Press, Oxford, pp. 227–245.

Antelman, S.M. and Chiodo, L.A. (1983) Amphetamine as a stressor. In: Creese, I. (ed.), *Stimulants: Neurochemical, Behavioural and Clinical Perspectives*. Raven Press, New York, pp. 269–290.

Antelman, S.M. and Chiodo, L.A. (1984) Stress: its effect on interactions among biogenic amines and role in the induction and treatment of disease. In: Iversen, L.L., Iversen, S.D. and Snyder, S.H. (eds), *Handbook of Psychopharmacology*. Plenum Press, New York, pp. 279–341.

Azrin, N.H. (1961) Time-out from positive reinforcement. *Science* 133, 382–383.

Brett, L. and Levine, S. (1979) Schedule-induced polydipsia suppresses pituitary–adrenal activity in rats. *Journal of Comparative Physiological Psychology* 93, 946–956.

Brett, L. and Levine, S. (1981) The pituitary–adrenal response to 'minimized' schedule-induced drinking. *Physiology and Behaviour* 16, 153–158.

Campbell, B.A. and Fibiger, H.C. (1971) Potentiation of amphetamine induced arousal by starvation. *Nature* 233, 424–5.

CEC (Commission of the European Communities) (1983) *Abnormal Behaviour in Farm Animals*. Commission of the European Communities, Brussels.

Cherek, D.R. and Brauchi, J.T. (1981) Schedule-induced cigarette smoking behaviour during fixed-interval monetary reinforced responding. In: Bradshaw, C.M., Szabadi, E. and Lowe, C.F. (eds), *Quantification of Steady-State Operant Behaviour*. Elsevier/North-Holland Biomedical Press, Amsterdam, pp. 389–392.

Costall, B. and Naylor, R.J. (1977) Mesolimbic and extrapyramidal sites for the mediation of stereotyped behaviour patterns and hyperactivity by amphetamine and apomorphine in the rat. In: Ellinwood, E.H. and Kilbey, M.M. (eds), *Cocaine and other Stimulants*. Plenum Press, New York, pp. 47–76.

Creese, I. and Iversen, S.D. (1975) The pharmacological and anatomical substrates of the amphetamine response in the rat. *Brain Research* 83, 419–436.

Cronin, G.M. and Wiepkema, P.R. (1984) An analysis of stereotyped behaviour in tethered sows. *Annales de Recherches Vétérinaires* 15, 263–270.

Cronin, G.M., Wiepkema, P.R. and Van Ree, J.M. (1985) Endogenous opioids are involved in abnormal stereotyped behaviours of tethered sows. *Neuropeptides* 6, 527–530.

Cronin, G.M., Wiepkema, P.R. and Van Ree, J.M. (1986) Endorphins are implicated in stereotypies of tethered sows. *Experientia* 4, 198–199.

Dantzer, R. (1986) Behavioural, physiological and functional aspects of stereotyped behaviour: a review and reinterpretation. *Journal of Animal Science* 62, 1776–1786.

Dantzer, R. (1989) Neuroendocrine correlates of control and coping. In: Steptoe, A. and Appels A. (eds), *Stress, Personal Control and Health*. Wiley, Chichester, pp. 277–294.

Dantzer, R. (1991) Stress, stereotypies and welfare. *Behavioural Processes* 25, 95–102.

Dantzer, R. and Mormède, P. (1981) Pituitary–adrenal correlates of adjunctive activities in pigs. *Hormones and Behaviour* 16, 78–92.

Dantzer, R. and Ollat, H. (1991) Alcoholism: a psychobiological perspective. *European Psychiatry* 6, 209–215.

Dantzer, R., Bluthé, R.M. and Tazi, A. (1986a) Stress-induced analgesia in pigs. *Annales de Recherches Vétérinaires* 17, 147–151.

Dantzer, R., Gonyou, H.W., Curtis, S.E. and Kelley, K.W. (1986b) Changes in serum cortisol reveal functional differences in frustration-induced chain chewing in pigs. *Physiology and Behaviour* 39, 775–777.

Dantzer, R., Terlouw, C., Mormède, P. and Le Moal, M. (1988a) Schedule-induced polydipsia experience decreases plasma corticosterone levels but increases plasma prolactin levels. *Physiology and Behaviour* 43, 275–279.

Dantzer, R., Terlouw, C., Tazi, A., Koolhas, J.M., Bohus, B., Koob, G.F. and Le Moal, M. (1988b) The propensity for schedule- induced polydipsia is related to differences in conditioned avoidance behaviour and in defense reaction in a defeat test. *Physiology and Behaviour* 43, 269–273.

Devenport, L.D. (1978) Schedule-induced polydipsia in rats: adrenocortical and hippocampal modulation. *Journal of Comparative and Physiological Psychology* 92, 651–660.

Falk, J.L. (1961) Production of polydipsia in normal rats by an intermittent food schedule. *Science* 133, 195–196.

Falk, J.L. (1966) The motivational properties of schedule induced polydipsia. *Journal of the Experimental Analysis of Behaviour* 9, 19–25.

Falk, J.L. (1971) The nature and determinants of adjunctive behaviour. *Physiology and Behaviour* 6, 577–588.

Falk, J.L. (1977) The origin and functions of adjunctive behaviour. *Animal Learning and Behaviour* 5, 325–335.

Fentress, J.C. (1983) Ethological models of hierarchy and patterning of species-specific behaviour. In: Satinoff, E. and Teitelbaum, P. (eds), *Handbook of Behavioural Neurobiology, Vol. 6*. Plenum Press, New York, pp. 185–234.

Fitzgerald, F.L. (1967) Effects of d-amphetamine upon behaviour of young chimpanzees reared under different conditions. In: Brill, H., Cole, J.O., Deniker, P.,

Hippius, H. and Bradley, P.B. (eds) *Proceedings of the Fifth International Congress of Neuropsychopharmacology*. Excerpta Medica, Amsterdam, pp. 1226–1237.

Freed, E.X. and Hymowitz, N. (1969) A fortuitous observation regarding 'psychogenic' polydipsia. *Psychological Review* 24, 224–226.

Gentry, W.D. (1968) Fixed-ratio schedule-induced aggression. *Journal of the Experimental Analysis of Behaviour* 11, 813–817.

Hennesssy, J.W. and Levine, S. (1979) Stress, arousal and the pituitary–adrenal system: a psychoendocrine model. In: Sprague, J. and Epstein, A. (eds), *Progress in Psychobiology and Physiological Psychology, Vol. 8*. Academic Press, New York, pp. 133–178.

Heybach, J.P. and Vernikos-Danellis, J. (1979a) Inhibition of adrenocorticotrophin secretion during deprivation-induced eating and drinking in rats. *Neuroendocrinology* 28, 329–338.

Heybach, J.P. and Vernikos–Danellis, J. (1979b) Inhibition of pituitary–adrenal response to stress during deprivation-induced feeding. *Endocrinology* 104, 967–973.

Honma, K., Honma, S., Hirai, T., Katsuno, Y. and Hiroshige, T. (1986) Food ingestion is more important to plasma corticosterone dynamics than water intake in rats under restricted daily feeding. *Physiology and Behaviour* 37, 791–795.

Hutchinson, R.R., Azrin, N.H. and Hunt, G.M. (1968) Attack produced by intermittent reinforcement of a concurrent operant response. *Journal of the Experimental Analysis of Behaviour* 11, 489–495.

Insel, T.R. (1988) Obsessive-compulsive disorder: new models. *Psychopharmacology Bulletin* 24, 365–369.

Jones, G.H., Mittleman, G. and Robbins, T.W. (1989) Attenuation of amphetamine-stereotypy by mesostriatal dopamine depletion enhances plasma corticosterone: implications for stereotypy as a coping response. *Behavioural and Neural Biology* 51, 80–91.

Kachanoff, R., Leveille, R., McClelland, J.P. and Wayner, M.J. (1973) Schedule induced behaviour in humans. *Physiology and Behaviour* 11, 395–398.

Kelly, P.H., Seviour, P. and Iversen, S.D. (1975) Amphetamine and apomorphine responses in the rat following 6-OHDA lesions of the nucleus acumbens septi and corpus striatum. *Brain Research* 94, 507–522.

Killeen, P. (1975) On the temporal control of behaviour. *Psychological Review* 82, 89–115.

Killeen, P.R., Hanson, S.J. and Osborne, S.R. (1978) Arousal: its genesis and manifestation as response rate. *Psychological Reviews* 85, 571–581.

Leshner, A.I. (1978) *An Introduction to Behavioural Endocrinology*. Oxford University Press, Oxford, p. 39.

Levine, R. and Levine, S. (1989) Role of the pituitary–adrenal hormones in the acquisition of schedule-induced polydipsia. *Behavioural Neuroscience* 103, 621–637.

Levine, S., Weinberg, J. and Ursin, H. (1978) Definition of the coping process and statement of the problem. In: Ursin, H., Baade, E. and Levine, S. (eds), *Psychobiology of Stress: A Study of Coping Men*. Academic Press, New York, pp. 3–21.

Levitsky, D. and Collier, G. (1968) Schedule-induced wheel running. *Physiology and Behaviour* 8, 571–573.

MacLennan, A.J. and Maier, S.F. (1983) Coping and stress-induced potentiation of stimulant stereotypy in the rat. *Science* 219, 1091–1093.

Mason, G.J. (1991) Stereotypies: a critical review. *Animal Behaviour* 41, 1015–1037.

Mendelson, J. and Chillag, D. (1970) Schedule-induced air licking in rats. *Physiology and Behaviour* 9, 535–537.

Meyer-Holzapfel, M. (1968) Abnormal behaviour in zoo animals. In: Fox, M. (ed.), *Abnormal Behaviour in Animals*. Saunders, London, pp. 476–503.

Mittleman, G. and Valenstein, E.S. (1984) Ingestive behaviour evoked by hypothalamic stimulation and schedule-induced polydipsia are related. *Science* 224, 415–417.

Mittleman, G., Castaneda, E., Robinson, T.E. and Valenstein, E.S. (1986) The propensity for nonregulatory ingestive behaviour is related to differences in dopamine systems: behavioural and biochemical evidence. *Behavioural Neuroscience* 100, 213–220.

Mittleman, G., Jones, G.H. and Robbins, T.W. (1988) The relationship between schedule-induced polydipsia and pituitary–adrenal activity: pharmacological manipulations and individual differences. *Behavioural Brain Research* 28, 315–324.

Mittleman, G., Brener, J. and Robbins, T.W. (1990) Physiological correlates of schedule-induced activities in rats. *American Journal of Physiology (Regulatory, Integrative and Comparative Physiology)* 259, R485–R491.

Mittleman, G., Jones, G.H. and Robbins, T.W. (1991) Sensitization of amphetamine-stereotypy reduces plasma corticosterone: implications for stereotypy as a coping response. *Behavioural and Neural Biology* 56, 170–182.

Mittleman, G., Blaha, C.D. and Phillips, A.G. (1992) Pituitary–adrenal and dopaminergic modulation of schedule-induced polydipsia: behavioural and neurochemical evidence. *Behavioral Neuroscience* 106, 408–420.

Moon, B.H., Feigenbaum, J.J., Carson, P.E. and Klawans, H.L. (1980) The role of dopaminergic mechanisms in naloxone-induced inhibition of apomorphine-induced stereotyped behaviour. *European Journal of Pharmacology* 61, 71–75.

Randrup, A. and Munkvad, I. (1967) Stereotyped behaviour produced by amphetamine in several animal species and man. *Psychopharmacologia* 11, 300–310.

Rapoport, J.L. and Wise, S.P. (1988) Obsessive-compulsive disorder: evidence for basal ganglia dysfunction. *Psychopharmacology Bulletin* 24, 380–384.

Ridley, R.M. and Baker, H.F. (1982) Stereotypy in monkeys and humans. *Psychological Medicine* 12, 61–72.

Riley, A.L. and Wetherington, C.L. (1987) The differential effects of naloxone hydrochloride on the acquisition and maintenance of schedule-induced polydipsia. *Pharmacology, Biochemistry and Behaviour* 26, 677–682.

Robbins, T.W., Mittleman, G., O'Brien, J. and Winn, P. (1990) The neuropsychological significance of stereotypy induced by stimulant drugs. In: Cooper, S.J. and Dourish, C.T. (eds), *Neurobiology of Stereotyped Behaviour*. Clarendon Press, Oxford, pp. 25–63.

Rushen, J., de Passillé A.M.B. and Schouten W. (1990) Stereotypic behaviour, endogenous opioids and postfeeding hypoalgesia in pigs. *Physiology and Behaviour* 48, 91–96.

Sahakian, B.J. and Robbins, T.W. (1975) Potentiation of locomotor activity and modification of stereotypy by starvation in apomorphine treated rats. *Neuropharmacology* 14, 251–257.

Sahakian, B.J., Robbins, T.W., Morgan, M.J. and Iversen, S.D. (1975). The effects of psychomotor stimulants on stereotypy and locomotor activity in socially deprived and control rats. *Brain Research* 84, 195–205.

Seguy, F. (1990) Polymorphisme de l'hyperthermie émotionnelle chez le rat: etude psychobiologique et pharmacologique. DEA Neurosciences, University of Bordeaux II.

Soussignan, R. and Koch, P. (1985) Rhythmical stereotypies (leg-swinging) associated with reductions in heart rate in normal school children. *Biological Psychology* 21, 161–167.

Swenson, R.M. and Vogel, W.H. (1983) Plasma cathecholamine and corticosterone as well as brain catecholamine changes during coping in rats exposed to stressful footshock. *Pharmacology, Biochemistry and Behaviour* 18, 689–694.

Taylor, D.B. and Lester, D. (1969) Schedule-induced nitrogen drinking in the rat. *Psychonomic Science* 1S, 17–18.

Tazi, A., Dantzer, R., Mormède, P. and Le Moal, M. (1986) Pituitary–adrenal correlates of schedule-induced polydipsia and wheel-running in rats. *Behavioural and Brain Research* 19, 249–256.

Tazi, A., Dantzer, R. and Le Moal, M. (1987) Prediction and control of food rewards modulate endogenous pain inhibitory systems. *Behavioural and Brain Research* 23, 197–204.

Teitelbaum, P. (1977) Levels of integration of the operant. In: Honig, W.K. and Staddon, J.E.R. (eds), *Handbook of Operant Behaviour*. Prentice-Hall, Englewood Cliffs, New Jersey, pp. 7–27.

Terlouw, E.M.C., Lawrence, A., Ladewig, J., de Passillé, A.M., Rushen, J. and Schouten, W.G.P. (1991) Relationship between plasma cortisol and stereotypic activities in pigs. *Behavioural Processes* 25, 133–153.

Thompson, D.M. (1964) Escape from S-delta associated with fixed-ratio reinforcement. *Journal of the Experimental Analysis of Behaviour* 8, 1–8.

Uvnas-Moberg, K. (1987) Gastrointestinal hormones and pathophysiology of functional gastrointestinal disorders. *Scandinavian Journal of Gastroenterology* 22, Suppl. 128, 138–146.

Wallace, M. and Singer, G. (1976) Schedule induced behaviour: a review of its generality, determinants and pharmacological data. *Pharmacology, Biochemistry and Behaviour* 8, 483–490.

Wallace, M., Singer, G., Finaly, J. and Gibson, S. (1983) The effect of 6-OHDA lesions of the nucleus accumbens septum on schedule-induced drinking, wheel running and corticosterone levels in the rat. *Pharmacology, Biochemistry and Behaviour* 18, 129–136.

Wiepkema, P.R., Van Hellemond, K.K., Roessingh, P. and Romberg, H. (1987) Behaviour and abosamal damage in individual veal calves. *Applied Animal Behaviour Science* 18, 257–268.

Wilkinson, C.W., Shinsako, J. and Dallman, M.F. (1982) Rapid decreases in adrenal and plasma corticosterone concentrations after drinking are not mediated by changes in plasma adrenocorticotropin concentration. *Endocrinology* 110, 1599–1606.

Wright, J.W. and Kelso, S.C. (1981) Adrenal demedullation suppresses schedule-induced polydipsia in rats. *Physiology and Behaviour* 26, 1–5.

Zanella, A.J., Broom, D.M. and Hunter, J. (1991) Changes in opioid receptors of sows in relation to housing, inactivity and stereotypies. In: Appleby, M.C., Horrell, R.I., Petherick, J.C. and Rutter, S.M. (eds), *Applied Animal Behaviour: Past, Present and Future*. Universities Federation for Animal Welfare, Potters Bar, UK, pp. 140–141.

Future Research Directions   8

FRANK O. ÖDBERG
State University of Ghent, Merelbeke, Belgium.

Editors' Introductory Notes:
As we move towards the summary of this book we have invited Ödberg to ponder on the future of stereotypy research. We asked Ödberg to set out his thoughts partly because he is one of the 'fathers' of stereotypy research but also because his interests encompass a broad spectrum of disciplines from applied ethology to cognitive psychology and neuropsychology. He is therefore rather uniquely capable of overviewing future prospects bearing in mind past successes (and failures).

Most importantly, all involved in stereotypy research should take comfort that he clearly still feels that there *is* a future and that there are many important questions still to be addressed. He reminds us that 'little is new under the sun', and that we must be good scholars of the past in order to see the future clearly. In the same breath he expresses concern about scientific 'sectarianism', and the apparent disinterest in understanding developments in other disciplines. The lessons of the past teach that ethology can learn from psychopharmacology and vice versa.

In the second part of the chapter Ödberg emphasizes the need to consider unifying characteristics of stereotypies, on the basis that a significant 'upsurge' in new ideas and approaches would result from serious attempts to 'model' the processes involved in stereotypies. He looks at various theoretical approaches from a range of disciplines that could serve as useful starting points.

Finally, he presents a summary of his suggestions for future research. In his last point he urges scientists involved in stereotypy research not to become repetitive themselves, but to keep an open (and flexible) mind for different explanations for the phenomenon.

The purpose of this chapter is to browse through various aspects of past and present research on stereotypies without any pretence of exhaustiveness in order to perform some brain-storming for future research.

Historical Perspective

In order to choose the best research strategies for the future, we have to know something about the past. Modesty is one of the things the past teaches us. One often thinks of a bright idea, for example, a new way of approaching a problem. When gathering some literature for the remarkable paper to be presented at the next conference or for the landmark article to be published, it is more than common to discover that somebody had the same idea years ago. Some might act as if they did not know of this previous work. In other instances, however, some fail to discover that the idea is not new for a variety of reasons:

1. They found the publication but read only the abstract or the paper itself superficially.
2. They never read the paper but took the reference from another article. (This can have two annoying consequences: conclusions supported by scanty data sometimes obtain the status of established knowledge and an erroneous citation is not corrected and others continue to wrongly cite it.)
3. One cannot read everything (fair enough!).
4. They did not find the publication because it appeared in a journal dealing with a different specialization in the behavioural sciences.
5. They simply never bothered to examine whether the subject had been studied previously and entered a new field with the heavy shoes of their complacency.
6. Most English speakers remain woefully ignorant of publications in other languages.

All these factors are in part responsible for the level of redundancy found in animal welfare research.

Let us consider the following points concerning stereotypies: their essential characteristics, the environmental causal situations and the motivational aspects, the internal neurophysiological and biochemical mechanisms, the phenomenon of emancipation, the problem of a putative function. All of these reviewed in this book have been dealt with by earlier authors, some even from the beginning of the century. Their work was sometimes anecdotal, but sometimes based on precise observations and counts. Sometimes the environment was changed in order to discover the external causal factor, but more often in relation to case studies than systematically planned experiments. Sometimes psychoanalysts used a kind of vague 'Deutung' (a German term meaning approximately 'interpretation'). Greenacre (1954), for example, explained head-banging in children by a 'need to establish a body reality in that particular area'. In any case they had to work without the technical tools we now have. They also lacked the feedback of the results of systematic behavioural and physiological research in various species, which later led to a more careful and sophisticated conceptual framework. However, scientific thinking usually develops that way. That is phenomena are classified into categories and processes within concepts as parsimoniously as possible, while one realizes later that reality is more complicated and concepts too monolithic. For instance, the chapter on motivation in this book clearly illustrates how careful one should be. On the whole, ideas about motivation and the mechanisms underlying stereotypies have balanced between the need for parsimony (the pursuit of the unifying theory) and warnings against the danger of mixing up different phenomena. Nevertheless, we must credit our senior colleagues for recognizing most key issues and often formulating the right questions long before they had the means to answer them.

The three 'classical' characteristics defining stereotypies, that is, relatively invariant pattern, regular repetition and apparent uselessness, are rather often associated with my name because I mentioned them in a previous paper (Ödberg, 1978). In fact, I do not deserve any credit for that as I only made a synthesis of what could be found in the ethological, psychopharmacological and neuropsychiatric literature in those days. Furthermore, other authors had also distilled these three criteria (among others, Fox, 1965; Hutt and Hutt, 1965; Keiper, 1969; Kiley-Worthington, 1977). Psychiatrists have often not been trained to record behaviour objectively so stereotypies are seldom well defined in the psychiatric literature. However, some authors did carry out relatively accurate clinical descriptions yielding (pre-ethological) definitions. For example in 1922, Kläsi published the following definition which I translate from German as literally as possible (I have split the text myself in order to emphasize the importance of each part). Stereotypies are: (i) expressions from the motor, verbal and mental field; (ii) which can be repeated; (iii) for a very long time by people; (iv) in the same form; (v) which became autonomous, completely separated from their origin; (vi) and which express no mood nor are adapted to any particular aim in objective reality. According to the French psychiatrist Guiraud (1950), the invariant motor pattern is due to neurophysiological mechanisms of learning, while the regular repetition is caused by a dysfunction of the central grey nuclei. These insights are remarkably inspiring for future modern neurophysiological research. The phenomenon of emancipation had already been recognized: see point (v) of Kläsi's (1922) definition, while Dromard and Abély (1916) think they have no 'ideo-affective' content. This reminds us of the recent suggestion by Dantzer (1986) that stereotypies could be pure motor automatisms without emotional implications.

The trouble was that these early authors had to find a way of delineating phenomenologically the concept in order to improve semantics before even knowing practically anything about stereotypies. Conflict-induced stereotypies had to be distinguished from, for example, those appearing after organic trauma or tumours, although it does not mean that understanding the one could not help understanding the other. However, definitions can handicap further research rather than support it. For example, one is often confronted with the problem of not possessing an adequate alternative terminology. This has happened with the concepts of 'psyche' and 'soma'. Man thinks in dichotomies. Descartes, among others, left us a nasty heritage by proposing a completely distinct nature for 'mind' and 'body' (see also Chapter 4). Now we are left without the right words to express modern views that recognize the unity between 'mind' and 'body' (Dantzer, 1989). This in turn also makes it difficult to think about reality. Luckily, our perception of the world possesses some independence from words, and experimental work can make progress aided by the most recent techniques, which in turn should help change our concepts and terminology. Meanwhile, we are still struggling with inadequate terms. Ideally we should try to find other words for phenomena encompassing stereotypies, which free us from some ideas associated with the old word, in order to plan our investigations in a fresh way.

The effect of various environmental factors had already been investigated before World War II in domestic and zoo animals (for a review see Meyer-Holzapfel, 1968). Such studies showed the importance of a detailed knowledge of the animal one is dealing with and its physiological and behavioural 'needs' (for warnings against a simplistic use of such a term, see Dawkins, 1983) in order to discover which part of the management or housing was responsible for the stereotypy. (It seems paradoxical, but the recent difficulties in discovering the origin of stereotypies in individuals of various pig breeds suggests how little we still know about the behaviour of a highly domesticated species.) These studies are a must for the 'pragmatic' ethologist who is only interested in the suppression of the appearance of stereotypies by improving housing conditions and management but they do not tell us a lot about the mechanisms underlying such stereotypies.

Many authors have endeavoured in the past to present a list of possible 'causes' for stereotypies. Often situations were described as different 'causes' without anyone realizing they contained common factors. Such initiatives could be useful in the management of a given species, but one must realize that the causes of stereotypies could be very numerous. Furthermore, many stereotypies could have a multifactorial origin, for example the summation of hunger and space restriction combined with given predispositions in a given individual. We have no certainty that common mechanisms are always responsible for the proper development of stereotypies, irrespective of the motivation involved. Consequently, 'causes' should be much more precisely identified than is usual ('absence of straw', 'boredom', 'lack of space') in order to be useful.

Such lists of causes could allow interesting comparisons in an evolutionary perspective. For example, which stereotypies can occur at which level of complexity of the brain? Which types of factors lead to stereotypies in which species? This would require a comparative approach with the framework of MacLean's 'triune brain' as background, taking into account the interactions between more primitive and recent structures (for a summary of MacLean's views, see Valzelli, 1980). This could add to evolutionary insight on how cognitive factors gradually interact with central self-organizational processes. With drug-induced stereotypies it is known that the higher the evolutionary status of the species, the more varied the motor patterns which can appear as stereotypies and the stronger individual differences are (Kjellberg and Randrup, 1972). This suggests an increasing interaction between neocortical structures, the limbic system and the striatum. The latter has been the favourite object of neurophysiological and biochemical studies. However, there are two experiments from the 1970s which I always keep in the back of my mind. Amphetamine could still induce stereotypies in thalamic rats (Huston and Borbély, 1974) and cats (Marcus and Villablanca, 1974) (i.e. in animals without neocortex, striatum and the largest part of the rhinencephalon). Eleven out of 20 thalamic rats showed stereotyped gnawing or head movements with sniffing. Interestingly enough, four out of the 11 rats had also shown these stereotypies spontaneously after the operation. Amphetamine could thus facilitate a behaviour, the threshold of which was lowered by the removal of higher brain centres. One thalamic cat still stereotyped after a complete section just above the quadrigeminal bodies of the mesencephalon and another one stopped doing so after a section at the level of the pons! These data suggest, in any case, that the striatum is not as essential as is often thought.

Perhaps I failed in understanding the origins of the neurophysiological, biochemical and neuropsychological approach, but my feeling is that the strong interest from those sciences for stereotypies since the end of the 1960s is due to availability of techniques, the commercial scramble for new psychopharmacological drugs, and the development of the classical 'model' of dopamine-agonist-induced stereotypies for schizophrenia (e.g. Randrup and Munkvad, 1967). Here again, one must regret the long-lasting lack of interaction between pharmacologists working with drug-induced stereotypies and ethologists working with conflict-induced ones. Pharmacological work is now increasingly being cited by the latter. The former have only just started to realize that the use of ethologically more meaningful situations could help them escape from the vicious circle in which psychopharmacology has been trapped for more than 20 years, methodologically and conceptually (see e.g. Olivier et al., 1991). In order to set up animal 'models' for human disorders, pharmacologists have sometimes fallen into tautological reasonings. For instance some avoidance behaviours were labelled as representing 'anxiety'; a drug which inhibited those behaviours was then considered to be anxiolytic. After the initial great discoveries of the neuroleptics, the benzodiazepines, the tricyclic antidepressive drugs and MAO-inhibitors, research yielded only variations of the same without any new breakthrough. The more recent development of 'etho-pharmacology' has already contributed to the discovery of substances with unexpected anxiolytic capacities. The rather unethological 'models' were in fact inhibiting further progress.

The current fashionable theme concerning the functional basis of stereotypies is also not new. Keiper (1969), Hutt and Hutt (1970) and Duncan and Wood-Gush (1972), wondered whether they reflected or regulated arousal. That is do they only express a given state or are they instrumental in lowering or increasing arousal? Berkson (1967) suspected it would be very difficult to distinguish experimentally between both as movements always imply proprioceptive stimulation. Some years ago, the 'anti-psychiatry' movement also followed a functionality reasoning. Some people cope with difficult life situations by developing a given psychosis according to their predispositions; do we have the right to bring them back to reality? The greater ability to escape into the comforting sphere of imagination or into psychosis is, as far as we know, an 'advantage' man has on other animals (Laborit, 1976, 1986). For some time a debate took place whether stereotypies produced stimulation in order to increase arousal (Morris, 1964; Keiper, 1970) or isolated the organism in order to decrease it (Fentress, 1976; Valenstein, 1976). Maybe this rather sterile debate would not have happened if one had considered that they could in fact do both; they could be part of a homeostatic mechanism, whatever the deviation from the norm. Kiley-Worthington (1977) expressed it in another way: '... a very low level of sensory input is as insupportable as a very high level'. Incidentally, the American experimental psychologists Maier and Ellen had already suggested in 1951 that anxiety reduction was responsible for 'fixated behaviour' elicited in the 'Maier paradigm' (see below), but they are seldom cited in the ethological literature.

This brings up the problem of interdisciplinarity again. It is striking how many different specializations within behavioural sciences (ethology, psychopharmacology, experimental psychology, behaviourism, developmental psychology, neuropsychology, psychiatry and more recently applied ethology) have all dealt with stereotypies or very relevant phenomena with few interactions between the separate disciplines. At the 1978 Madrid conference on Ethology Applied to Zootechnics, I tried to stimulate interdisciplinary contact by bringing together in a round-table a pharmacologist (Axel Randrup), an experimental psychologist (Bob Feldman), a biology-trained ethologist (Ron Keiper), a veterinary surgeon (Roger Ewbank) and a psychology trained ethologist (myself). I must confess it enriched us more than it resulted in widespread interest. There has also been quite a lot of literature on stereotypies written by developmental psychologists working either with primates (the 'isolation syndrome') or with children, but it would take too much space to review it here.

Besides the blinkers of academic barriers, other reasons for the lack of interaction could be that the same type of phenomenon was described and analysed under a different name, or that different behaviours were wrongly labelled as stereotypies. The most messy situation is found in psychiatry where tics, mannerisms, perseverations and compulsions were sometimes called stereotypies. 'Adjunctive behaviours' (e.g. Wayner, 1970, 1974; Falk, 1971; Singer *et al.*, 1974) described by more behaviouristically oriented psychologists resemble displacement activities whose repetition often represents the initial stages of development of stereotypies. Marx (1972), in similar causal situations, described 'unconditioned chewing responses to frustration' in rats. Another class of behaviours which are also relevant are the 'superstitious behaviours' (Davis and Hubbard, 1972). Both are sometimes referred to by students of stereotypy.

An interesting phenomenon which did not attract the attention it deserved is 'fixated behaviour' occurring in 'Maier's paradigm' (Maier, 1949; Yates, 1962; Feldman, 1978). The Lashley jumping stand, originally developed to study visual discrimination, consists of a vertical panel with two small doors covered by, for example, a symbol such as a circle or triangle. The positions of the symbols change randomly. A rat sits on a stand some distance from the panel and has to learn that one of the symbols is the right choice (i.e. that door can be pushed open by jumping on it and gives access to a reward behind it). A wrong choice means bumping into a closed door and a fall into a net underneath. Maier conceived the following procedure in three phases:

1. 'Shaping': the animal learns the right choice.
2. In a second phase the problem is made unsolvable (i.e. there is no rule as both symbols are rewarded and punished an equal number of times according to a random sequence). The rats are obliged to jump by an electric shock or a blow to the tail, or by being pushed. Animals react in one of the following ways:
   (a) 'Position-stereotyped' (80% of stereotyping animals): the rat jumps systematically to the left or to the right door, whatever symbol it carries;
   (b) 'Symbol-stereotyped' (20% of stereotyping animals) the rat jumps systematically to one of the symbols, wherever it stands;
   (c) Escape reactions: the rat tries to escape from the situation by jumping, for example, over the panel or directly into the net.
3. In a third phase the initial rule is reintroduced, that is a given symbol represents again the right choice. Most interesting is that 75% of the position-stereotyped rats and 25% of the symbol-stereotyped ones persevere in their stereotyped choice even when the problem can again be solved. Maier called these 'fixated' when this persistence occurred over 200 trials (10 per day), as he found out that longer testing did not reveal any change.

This paradigm is interesting as it is more the choice as such which is stereotyped than a given motor pattern which is automatically repeated. The choice occurs as reaction to a given stimulus (i.e. the obligation to jump) which is not permanently present in the life of the animal. This could help us in understanding the mechanisms of cognitive aspects of the initial stages of stereotypies. It could also be closer to mentally recurring ideas in psychiatric patients than motor stereotypies.

Unfortunately, terminological differences were not the only reason for mutual ignorance. The boundaries between 'fundamental' and 'applied' science are often vague and sometimes even undesirable. The field of applied ethology in particular has suffered a lot from the lack of contact with more 'fundamental' behavioural sciences. Especially concerning abnormal behaviours, in which applied ethology was interested because of its welfare implications, many redundant words and useless neologisms were told and published. Behavioural reactions of production animals in conflict situations can be described by the existing concepts and terminology of ethology and experimental psychology without having to create new ones. Culmination of sterile intellectual isolation threatened when, following a round-table on 'The nomenclature of applied ethology' at a conference in 1978 (see the *Proceedings of the 1st World Congress on Ethology Applied to Zootechnics*, 1978), it was suggested that a terminology proper to applied ethology be developed. This luckily produced quite radical negative reactions from ethologists themselves. This would have even more isolated from its roots a young scientific specialty which was barely able to stand independently on its feet.

On the whole, the main difference between recent stereotypy research and the past is twofold. First, competent interdisciplinary approaches are fortunately increasing. However, a lot of bridges will still have to be built. It is striking that the literature cited in most chapters of the very interesting book *Neurobiology of Stereotyped Behaviour* (edited by Cooper and Dourish) which appeared in 1990, is completely different from that cited by people involved with farm animals and welfare. This means that no specialty is free from blinkers. Efforts will have to be made by all sides. Basic training in particular sciences will probably remain indispensable in order to develop knowledgeable scientists with a critical mind. On the other side, the encyclopedical ideal of the renaissance cannot be realized due to the tremendous extension of knowledge. The answer lies in the education of scientists with an 'interdisciplinary aperture'. The readiness to go and ask for the advice of a specialist from another science should become an elementary attitude (Ödberg, 1991). Secondly, improvement of available technical tools has refined the analyses and extended the possibilities (e.g. telemetrical monitoring of physiological parameters, computerized event-recording, more sensitive biochemical assays, intra-cerebral microdialysis, positon cameras, etc.).

Attempts at a Unifying Theoretical Framework – A Critical Evaluation

Ideas written down concisely are apparently less likely to be noticed. One of the first trials for a unifying theoretical framework could be found in my short six pages of 1978. It will always remain frustrating for me that this paper is still often cited for its definition of stereotypies and not because of its synthetical presentation of most key questions integrated into one framework!

The best way to put some structure into this discussion is again to start from such a framework, but a slightly updated one. Parts of the rationale which follows can be found in Ödberg (1989), but that publication laid the emphasis on levels of adaptation. Let us use the cybernetical model which was introduced in animal welfare by Wiepkema (1983). A mismatch between 'Sollwert' (the norm) and 'Istwert' (the actual situation) results in some action of the organism in order to reduce the mismatch. If that action fails, it is labelled as inefficient and another one can be tried out (Gray, 1982). When no action succeeds, the degree of importance of the norm will determine whether the organism can shift its attention (and this can imply cognitive factors but not necessarily) to other regulatory processes. Alternatively, on the cognitive level the perceptual norm could be changed in order to match it with the situation. This is the 'sour grapes' rationalization and human psychologists have built whole theories around the avoidance of incoherence (e.g. Festinger's (1964) cognitive dissonance). When reduction of that particular mismatch is more important than others, the same action will be repeated anyway by some individuals although it does not solve that mismatch.

At this level of the discussion we can identify two questions. First, which mismatches induce stereotypies? This is a question about the 'causal situations'. It will depend on all the factors determining norms at a given age and status of an individual of a given species. I would like to restate the difference between what I used to call the 'original causal factor(s)', (i.e. that particular motivation, or combination of different ones which could not be fulfilled and which gave rise to a particular stereotypy), and the 'modulating factor(s)' (i.e. other stimuli which increase or decrease the stereotypy appearance). For example, Fentress (1976) was able to modulate the intensity of stereotypies in voles by noises or human presence. These stereotypies stopped when the stimulation was too high, but resumed with a 'rebound-effect' afterwards. Rattling with a ballpen over the metal lid of the cage of the voles in my colony increases the chances of observing stereotypies soon afterwards, while doing so while the animal is performing stereotypies will result in temporary inhibition. Stereotypy frequency in chimpanzees was increased by the emission of a loud noise, by hunger (Berkson and Mason, 1964) or by presentation of complex strange moving objects (Berkson et al., 1963). Loud noise had a similar effect in mentally deficient humans (Levitt and Kaufmann, 1965). Some animals increase their stereotypy frequency when feeding time approaches (e.g. horses and hyenas, Holzpfel, 1938, pigs, Fraser, 1975) or when defecating (horses: personal observation). As pointed out in Chapter 3, one is dealing here with the wider problem of general arousal versus specific stimulation. One could object to the notion of 'modulating factor' that many stereotypies have a multifactorial origin and that the supposed 'modulator' stimulus in fact belongs to one of the motivational systems involved in the original causal conflict. In many cases it will be very difficult to prove or to invalidate the hypothesis, and one will often have to resort to probabilities. For instance, 'enrichment' (twigs, partitions) of the cage inhibits the development of stereotypies in voles (Ödberg, 1987a). Among other factors (e.g. gnawing, nest-building, wall-seeking), 'enrichment' could contain the possibility to hide from fearful stimuli, such as human presence. In such a case, one could argue that the human presence does not influence stereotypy frequency by changing the general arousal level but rather increases the original conflict, due to the impossibility of hiding in that barren cage. On the other hand, although one cannot be absolutely sure, it seems rather unlikely that weaving in horses finds its origin in a thwarting of defecation. In any case, the theoretical possibility of a 'modulating factor' should never be discarded.

From the methodological point of view it is important to realize that once an older stereotypy is emancipated from the original causal situation, variation of the causal factor(s) could only result in effects similar to those of a 'modulating factor', which will complicate further the task of our pragmatic colleague in his endeavours to develop better husbandry systems. The importance of developmental studies should be stressed over and over again, not only for the study of causality but also for that of mechanisms. It is very important to know which developmental stage the stereotypies one is dealing with are in (see Chapter 2). Their neurobiochemical basis changes and the effect of drugs can vary. Duncan and Wood-Gush (1974) were able to inhibit the development of stereotypies in chickens with a rauwolfia-derivate when given during the initial conflict period (inaccessible food); the drug had no effect once stereotypies had developed. Feldman (1962, 1964, 1968) found the same using chlordiazepoxide and diazepam in Maier's paradigm. Voles bred in barren laboratory cages start stereotyping between the age of 20 and 30 days. Naloxone inhibits these stereotypies up to the age of 4 months, after which its effect disappears, while haloperidol conserves this effect (Kennes et al., 1988). It has been demonstrated empirically that certain individuals do not stop stereotyping when removed from the situation in which the stereotypy developed. It would be surprising if no neurological changes accompanied such a phenomenon. Interestingly, there seems to be a correlation between the age at which individual voles start showing emancipation (Cooper and Ödberg, 1991) and the end of naloxone inhibition found by Kennes et al. (1988) Ideally speaking, non-stereotyping individuals should be introduced into different environment(s) and the frequency of appearance of stereotypies followed and compared (e.g. Ödberg, 1987a; Schouten and Wiepkema, 1991; Terlouw et al., 1992). The second position that can be identified is: what determines which motor pattern becomes stereotyped? This question has been addressed elsewhere by several authors (e.g. Ödberg, 1978; Mason, 1991) and will not be dealt with here.

Let us take up the thread of our framework again. Why do some individuals repeat a given action instead of switching to another one? Various theories have tried to explain this repetition. Before reviewing these briefly, I would like to point out that very little has been done up to now to understand the animal's perception of the mismatch. Cognitive processing could result in 'anxiety' or 'frustration', with an increase of 'arousal'. In any case this will be a negative experience which should be avoided. In 1978, reacting against the 'lists of causes', I considered frustration to be the common denominator of all conflict-induced stereotypies (see also Ödberg, 1987b). However, we are still unsure whether the organism is always treating the matter on a conscious level. A homeostatic regulation of arousal could be performed unconsciously like many regulatory actions in the body. On the other hand, stereotypies which at first sight seem to be purely automatic consequences of a direct pharmacological stimulation of some parts of the striatum have been demonstrated to be influenced by cognitive elements. For example, novelty, social contact and past experiences can all influence the effect of amphetamine (see Chapter 6).

Furthermore, the target of amphetamine itself could be cognitive systems instead of motor ones. An experiment by Ridley *et al.* (1981) suggested that the perseverative effect of amphetamine could be due to an inability to alter the reward association of stimuli (a lack of cognitive flexibility) rather than an inability to break the motor habit of responding to one stimulus. More attention to cognitive approaches could throw a new light on the study of stereotypies.

Repetition has been explained in various ways:

1. The most simple explanation is that as stereotypies usually appear in situations with little variation, the organism keeps reacting to the few key stimuli for the motivation concerned which are still present (see Chapter 3).
2. By Dantzer's (1986, 1991) theory of sensitization and self-organization (see Chapter 3).
3. Stereotypies have also been attributed a homeostatic function (the 'coping' theory). Several studies have found stereotypies and schedule-induced behaviours to be associated with lowered cortisol/corticosterone levels, absence of gastric ulcers, lowered heart rate, EEG waves typical for drowsiness, etc. (see Chapters 5 and 7). Studies which are not often cited are those by Sørensen and Randrup (1986), Sørensen (1987) and Randrup *et al.* (1988). They found that in restricted cage environments laboratory-born voles develop more locomotor stereotypies while wild-caught ones develop polydipsia usually leading to death. They conclude that stereotypies may function as a survival strategy in an unfavourable environment. However, one could also suggest that voles which managed to survive in laboratory cages were those (or their descendants) with a propensity to develop stereotypies, while a direct causal relation between the stereotypy should not be taken for granted. Another alternative explanation is the following: if we consider polydipsia to be a stereotypy too, the main difference between the two populations consists of the type of stereotypy which they developed. It is a matter of chance that the one stereotypy has lethal consequences while the other has not.

Unfortunately, we know next to nothing concerning the exact nature of the relation between the performance of stereotypies as such and their supposed advantages.

First, the relationship could be correlational instead of causal. Stereotypies could be only part of a wider coping mechanism with no proper effect on the observed parameters, or could theoretically even be mere 'side-effects' without any role (Chapter 6). In this connection one should welcome the recent work of Bohus and Koolhaas on different 'coping strategies' among the same species, that is the so-called 'active' and 'passive' copers. One should warn, however, that it would be somewhat naïve to consider the stereotyping animals as the copers and the non-stereotyping animals as the unlucky ones. To begin with, the latter could be those that do not even experience the environment as negative. Furthermore, if individuals are equipped with either the tendency to try to cope actively or passively, this means that *both* types could be more or less successful. The methodological consequences are that, if one wants to explore the coping function hypothesis of stereotypies, one should not compare stereotyping with non-stereotyping animals. One should follow the chosen parameters in each individual and see whether differences eventually appear within each group (in terms of costs and benefits). A comparison between stereotypers and non-stereotypers remains meaningful if one desires to explore differences between them as such, without considering the coping function hypothesis. These differences could partly find their origin in, for example, different predispositions for activation of particular neurotransmitter systems. Cools *et al.* (1990) succeeded in selecting for apomorphine-susceptible and non-susceptible rats. However, one can work the other way round. Once individual differences in stereotypy levels have been determined, one can investigate in what other behavioural and physiological aspects these animals differ. For example, interestingly, Terlouw *et al.* (1992) found differences in behavioural responses to amphetamine according to level of environmentally induced stereotypy in pigs. These seem to be due to constitutional differences in the catecholaminergic system rather than to receptor sensitivity changes induced by the performance of the stereotypies themselves. Gradual harvesting of such data could then lead to the establishment of certain constitutional typologies. That type of research is also being undertaken concerning schedule-induced polydipsia and propensity to eat and drink in response to electrical stimulation of the lateral hypothalamus (see Dantzer, 1991, Chapter 7). Of course, even if a given constitutional set-up is responsible for the propensity to develop stereotypies, it does not enable us to decide about the homeostatic function. Furthermore, stereotypies could be indirectly functional, contributing to support other unsuspected mechanisms within the global strategy (e.g. by activating certain brain parts through proprioceptive stimulation, which in turn fulfil a proper role).

Second, the putative causal relation could be reversed. For example,

Redbo et al. (1992) found a relation between high milk yield and stereotypies in cows. They suggest that as high-yielding cows often fail to maintain their energy balance, that factor could contribute together with the physical restraint of tethering to the suboptimal environment and hence increase the probability of stereotypies.

The main message seems to be to refrain from jumping to conclusions in either way but to proceed with an open mind.

4. The final hypothesis is that stereotypies are rewarding through activation of endogenous opioids (EOPS) (see also Chapters 5, 6 and 7). In this part I will only discuss briefly some recent additional data. EOPS have been demonstrated to be involved in the initial stages of development of stereotypies. This, however, does not imply that their actual level is the direct result of the performance of stereotypies, although it is known that self-narcotization is possible (Watkins and Mayer, 1982). Such a relation has been shown in one automutilating Lesh-Nyhan patient (Richardson and Zaleski, 1986; Richardson et al., 1986). Infusion of naloxone induced an initial temporary increase of mutilations followed by a disappearance. The same was observed in rats which increased pushing for amphetamine self-administration after pimozide (Yokel and Wise, 1975). Our naloxone study (Kennes et al., 1988) used automatically recorded daily stereotypy averages and did not check for acute effects on a short-term basis. This is very difficult in voles because of their multi-phasic activity rhythm. A student of mine (I. Vandebroek, unpublished BSc. Thesis, 1991), recently observed the acute effect of three naloxone doses. Unfortunately, few data could be obtained because of the methodological difficulties mentioned above. In any case, no distinct increase of stereotypies following naloxone was observed. We cannot be certain as long as this experiment is not repeated, however the indication is that the previous results of 1988 could be interpreted as if EOPS sensitized dopamine receptors during the first stages, facilitating self-organization, thus supporting Dantzer's hypothesis (1986, 1991).

Playing heuristically with models is a game that can lead to new ideas for experiments. It is surprising how few authors have tried to fill the causal gap between stereotypies and their putative advantages by using the hypothesis of the 'behavioural inhibition system' (BIS) described by Gray (1982, 1984) and Laborit (1986). The BIS is activated when an important mismatch is detected or when something aversive occurs (novelty, frustrative non-reward, punishment). Execution of higher-level planning is interrupted, the actual motor programme is labelled as faulty or inefficient, arousal increases and alternatives are explored (increased attention). If the organism has lost control, that is when no action can reduce the mismatch related to an important norm, most actions are inhibited and the organism enters what is commonly called the 'conservation-withdrawal' state with stimulation of the pituitary–adrenal axis (see various chapters of the Ciba Foundation Symposium 8, 1972 and Henry and Stephens, 1977). According to endocrinologists there seems to be a positive feedback between glucocorticoids and the stimulation of the BIS (Laborit, 1986, p. 143). The only way to escape from that vicious circle seems to be to act anyway. Action as such seems to be important. Rats submitted to repeated unavoidable electric shocks show less ACTH increase (Conner et al., 1971) and do not develop a stable high blood pressure (Kunz et al., 1974) when they have the opportunity to attack a conspecific. It was later found that behaviours other than aggressive ones had the same effect. Hence it could be that in unescapable conflict situations the message is 'do something, whatever it is'. Certain behaviours could become rewarded that way. Applied to stereotypies, this model does not account well up to now for their very high repetitiveness but it could add to the understanding of their eventual coping function in those individuals which tend to react actively.

However, going a step further into the neuropsychology of the septo-hippocampal system and the prelimbic cortex could indicate to us new ways to investigate the problem of repetitivity. The prelimbic cortex seems to be involved in the shifting of cognitive strategies (Brito and Brito, 1990). The hippocampus is considered to be an interface between cognition and emotions (Gray, 1984) and the nucleus accumbens to be an interface between emotions and motor programmes, especially responsible for switching to cue-directed behaviours (van den Bos and Cools, 1989). The hippocampus seems to play an important role as a comparator between actual and expected stimuli and its integrity and maturity affect the inhibition of useless actions (for a review see Blozovski, 1986). Edelman (1989) extends even this comparative role in order to account for the global functioning of the central nervous system in terms of 'self' and 'non-self'. Could it be that some dysfunctions occur in these structures when an organism *starts* to repeat a given motor pattern although it should have been tagged as inefficient? Hippocampal functioning can be investigated for instance through spontaneous alternation behaviour (Dember and Richman, 1989).

Another theoretical model worth examining is that of 'deterministic chaos' which is increasingly being applied to biology (May, 1991). Behaviour fluctuates between different degrees of probability. Some sequences can be highly predictable and efficient in relatively unchanging environments, others are more stochastic and are more efficient in changing situations. Patterns that are too deterministic could result in the behavioural system getting stuck. Some 'noise' is necessary so that self-organization processes remain flexible (Nicolis and Prigogine, 1977). Concerning stereotypies we are dealing with the problem that behavioural variability decreases spectacularly in certain situations. Maybe the equations of the mathematical model could help us understand why they are not temporary states? Are stereotypies the result of a 'catastrophic' process leading, through a self-organization mechanism, to a state of no-return? In that case they could be a kind of non-functional 'by-product' of an eventually successful homeostatic process. An organism can sacrifice functions at some levels in order to save the whole structure (Laborit, 1986).

Summary of Suggestions for Future Research

1. Carry out as far as possible *developmental* studies, trying to *induce* stereotypies and investigating what *changes* in the organism.
2. Study *individual differences* (high/low, stereotypers/non-stereotypers, between stereotypers, between non-stereotypers), investigating in which aspects they differ *other than in performances of stereotyping*.
3. Use increasingly *interdisciplinary approaches*, especially neuropsychological and biochemical ones, with more attention on cognitive processes.
4. Keep an open mind for different hypotheses.

Acknowledgements

Thanks are due to U.A. Luescher and to the editors for their constructive comments.

References

Berkson, G. (1967) Abnormal stereotyped motor acts. In: Zubin, J. and Hunt, H.F. (eds), *Comparative Psychopathology. Animal and Human*. Grune and Stratton, New York, pp. 76–94.
Berkson, G. and Mason, W.A. (1964) Stereotyped movements of mental defectives. IV. The effect of toys and the character of the acts. *American Journal of Mental Deficiency* 68, 511–524.
Berkson, G., Mason, W.A. and Saxon, S.V. (1963) Situation and stimulus effects on stereotyped behaviours of chimpanzees. *Journal of Comparative and Physiological Psychology* 56, 786–792.
Blozovski, D. (1986) L'hippocampe et le comportement. *La Recherche* 17, 330–337.
Brito, G.N.O. and Brito, L.S.O. (1990) Septohippocampal system and the prelimbic sector of frontal cortex: a neuropsychological battery analysis in the rat. *Behavioural Brain Research* 36, 127–146.
Ciba Foundation Symposium (1972) Vol. 8 (new series), *Physiology, Emotions and Psychosomatic Illness*. Elsevier, Amsterdam.
Conner, R.L., Vernikos-Danellis, J. and Levine, S. (1971) Stress, fighting and neuroendocrine function. *Nature* 234, 564–566.
Cools, A.R., Brachten, R., Heeren, D., Willemen, A. and Ellenbroek, B. (1990) Search after neurobiological profile of individual-specific features of Wistar rats. *Brain Research Bulletin* 24, 49–69.
Cooper, S.J. and Dourish, C.T. (eds) (1990) *Neurobiology of Stereotyped Behaviour*. Clarendon Press, Oxford.

Cooper, S.J. and Ödberg, F.O. (1991) The emancipation of stereotypies with age (abstract). In: Appleby, M.C., Horrell, R.I., Petherick, J.C. and Rutten, S.M. (eds), *Applied Animal Behaviour: Past, Present and Future*. UFAW, Potters Bar, UK, p. 142.

Dantzer, R. (1986) Behavioural, physiological, and functional aspects of stereotyped behaviour: a review and a re-interpretation. *Journal of Animal Science* 62, 1776–1786.

Dantzer, R. (1989) *L'Illusion Psychosomatique*. Editions Odile Jacob, Paris.

Dantzer, R. (1991) Stress, stereotypies and welfare. *Behavioural Processes* 25, 95–102.

Davis, H. and Hubbard, J. (1972) An analysis of superstitious behaviour in the rat. *Behaviour* 43, 1–12.

Dawkins, M.S. (1983) Battery hens name their price: consumer demand theory and the measurement of ethological 'needs'. *Animal Behaviour* 31, 1195–1205.

Dember, W.N. and Richman, C.L. (1989) *Spontaneous Alternation Behaviour*. Springer-Verlag, New York.

Dromard and Abély (1916), cited by Michaux, L. (1965) *Psychiatrie*. Flammarion, Paris.

Duncan, I.J.H. and Wood-Gush, D.G.M. (1972) Thwarting of feeding behaviour in the domestic fowl. *Animal Behaviour* 20, 444–451.

Duncan, I.J.H. and Wood-Gush, D.G.M. (1974) The effect of a rauwolfia tranquilliser on stereotyped movements in frustrated domestic fowl. *Applied Animal Ethology* 1, 67–76.

Edelman, G. (1989) *The Remembered Present: A Biological Theory of Consciousness*, Basic Books, cited by Changeux, J.P. (1992) Les neurones de la raison. *La Recherche* 244, 704–713.

Falk, J.L. (1971) Production of polydipsia in normal rats by an intermittent food schedule. *Science* 133, 195–196.

Feldman, R.S. (1962) The prevention of fixations with chlordiazepoxide. *Journal of Neuropsychiatry* 3, 254–259.

Feldman, R.S. (1964) Further studies on assay and testing on fixation-preventing psychotropic drugs. *Psychopharmacologia* 6, 130–142.

Feldman, R.S. (1968) The mechanism of fixation prevention and 'dissociation' learning with chlordiazepoxide. *Psychopharmacologia* 12, 384–399.

Feldman, R.S. (1978) Environmental and physiological determinants of fixated behaviour in mammals. In: *Proceedings of the 1st World Congress on Ethology Applied to Zootechnics*. Editorial Garsi, pp. 487–493.

Fentress, J.C. (1976). Dynamic boundaries of patterned behaviour interaction and self-organization. In: Bateson, P.P.G. and Hinde, R.A. (eds), *Growing Points in Ethology*. Cambridge University Press, Cambridge, pp. 135–169.

Festinger, L. (1964) *Conflict, Decision and Dissonance*. Stanford University Press, Stanford.

Fox, M.W. (1965) Environmental factors influencing stereotyped and allelomimetic behaviour in animals. *Laboratory Animal Care* 15, 363–370.

Fraser, D. (1975) The effect of straw on the behaviour of sows in tether stalls. *Animal Production* 21, 59–68.

Gray, J.A. (1982) *The Neuropsychology of Anxiety; an Enquiry into the Functions of the Septo-hippocampal System*. Clarendon Press, Oxford/Oxford University Press, New York.

Gray, J.A. (1984) The hippocampus as an interface between cognition and emotion. In: Roitblat, H.L., Bever, T.G. and Terrace, H.S. (eds), *Animal Cognition*. Lawrence Erlbaum, London, pp. 607–626.

Greenacre, P. (1954) Problems of infantile neurosis: a discussion. *The Pyschoanalytical Study of the Child* 9, 16–71.

Guiraud, P. (1950) *Psychiatrie Générale*. Le Francois, Paris.

Henry, J.P. and Stephens, P.M. (1977) *Stress, Health and the Social Environment*. Springer-Verlag, New York.

Holzapfel, M. (1938) Über Bewegungsstereotypien bei gehaltenen Säugern. I. Mitteilung: Bewegungsstereotypien bei Caniden und Hyaena. II. Mitteilung: Das Weben der Pferde. *Zeitschrift für Tierpsychologie* 2, 46–71.

Huston, J.P. and Borbély, A.A. (1974) The thalamic rat: general behaviour, operant learning with rewarding hypothalamic stimulation, and effects of amphetamine. *Physiology and Behaviour* 12, 433–448.

Hutt, C. and Hutt, S.J. (1965) Effect of environment complexity upon stereotyped behaviours in children. *Animal Behaviour* 13, 1–4.

Hutt, C, and Hutt, S.J. (1970) Stereotypies and their relation to arousal: a study of autistic children. In: Hutt, S.J. and Hutt, C. (eds), *Behaviour Studies in Psychiatry*. Pergamon Press, Oxford, pp. 175–204.

Keiper, R.R. (1969) Causal factors of stereotypies in caged birds. *Animal Behaviour* 17, 114–119.

Keiper, R.R. (1970) Studies of stereotypy function in the canary (*Serinus canarius*). *Animal Behaviour* 18, 353–357.

Kennes, D., Ödberg, F.O., Bouquet, Y. and De Rycke, P.H. (1988) Changes in naloxone and haloperidol effects during the development of captivity-induced jumping stereotypy in bank voles. *European Journal of Pharmacology* 153, 19–24.

Kiley-Worthington, M. (1977) *Behavioural Problems of Farm Animals*. Oriel Press, Stocksfield.

Kjellberg, B. and Randrup, A. (1972) Stereotypy with selective stimulation of certain items of behaviour observed in amphetamine-treated monkey (*Cercopithecus*). *Pharmakopsychiatrie* 5, 1–12.

Kläsi, J. (1922) *Über die Bedeutung und Entstehung der Stereotypien*. Karger, Berlin.

Kunz, E., Valette, N. and Laborit, H. (1974) Rôle antagoniste de l'activité motrice d'évitement ou de lutte à l'égard de l'hypertension artérielle chronique provoquée chez le rat par l'application journalière d'un choc électrique plantaire. *Agressologie* 15, 333–339.

Laborit, H. (1976) *Eloge de la Fuite*. Gallimard, Paris.

Laborit, H. (1986) *L'Inhibition de l'Action. Biologie Comportementale et Physio-Pathologie*. Masson, Paris.

Levitt, H. and Kaufmann, M.E. (1965) Sound induced drive and stereotyped behaviour in mental defectives. *American Journal of Mental Deficiency* 69, 729–734.

Maier, N.R.F. (1949) *Frustration. The Study of Behaviour without a Goal*. McGraw-Hill, New York.

Maier, N.R.F. and Ellen, P. (1951) Can the anxiety-reduction theory explain abnormal fixations? *Psychological Review* 58, 435–445.

Marcus, R.J. and Villablanca, J.R. (1974) Is the striatum needed for amphetamine induced stereotyped behaviour? *Proceedings of the Western Pharmacological Society* 17, 219–222.

Marx, M.H. (1972) Unconditioned chewing response in the rat as a measure of reaction to frustration. *Psychological Report* 30, 613–614.
Mason, G.J. (1991) Stereotypies: a critical review. *Animal Behaviour* 41, 1015–1037.
May, R.M. (1991) Le chaos en biologie. *La Recherche* 22, 588–598.
Meyer-Holzapfel, M. (1968) Abnormal behavior in zoo animals. In: Fox, M.W. (ed.), *Abnormal Behavior in Animals*. W.B. Saunders, Philadelphia, pp. 476–503.
Morris, D. (1964) Responses of animals to a restricted environment. In: Edholm, O.G. (ed.), *The Biology of Survival*. Symposium of the Zoological Society of London 11:13, pp. 99–118.
Nicolis, G. and Prigogine, I. (1977) *Self-organization in Non-equilibrium Systems*. Wiley, New York.
Ödberg, F.O. (1978) Abnormal behaviours (stereotypies). In: *Proceedings of the 1st World Congress on Ethology Applied to Zootechnics*. Editorial Garsi, pp. 475–480.
Ödberg, F.O. (1987a) The influence of cage size and environmental enrichment on the development of stereotypies in bank voles (*Clethrionomys glareolus*). *Behavioural Processes* 14, 155–173.
Ödberg, F.O. (1987b) Behavioural responses to stress in farm animals. In: Wiepkema, P.R. and van Adrichem, P.W.M. (eds), *Biology of Stress in Farm Animals: An Integrative Approach*. Martinus Nijhoff, Dordrecht, pp. 135–150.
Ödberg, F.O. (1989) Behavioural coping in chronic stress conditions. In: Blanchard, R.J., Brain, P.F., Blanchard, D.C. and Parmigiani, S. (eds), *Ethoexperimental Approaches of the Study of Behavior*. Kluwer Academic Press, Dordrecht, pp. 229–238.
Ödberg, F.O. (1991) Introduction to Chapter IV. Clinical Psychobiology. In: Archer, T. and Hansen, S. (eds), *Behavioural Biology: Neuroendocrine Axis*. Lawrence Erlbaum Associates, Hillsdale, New Jersey pp. 167–171.
Olivier, B., Mos, J. and Slangen, J.L. (eds) (1991) *Animal Models in Psychopharmacology*. Birkhäuser Verlag, Basel.
Proceedings of the 1st Congress on Ethology Applied to Zootechnics. Madrid (1978), Editorial Garsi, pp. 535–537.
Randrup, A. and Munkvad, I. (1967) Stereotyped activities produced by amphetamine in several animal species and man. *Psychopharmacologia* 11, 300–310.
Randrup, A., Sørensen, G. and Kobayashi, M. (1988) Stereotyped behaviour in animals induced by stimulant drugs or by a restricted cage environment: relation to disintegrated behaviour, brain dopamine and psychiatric disease. *Japanese Journal of Psychopharmacology* 8, 313–327.
Redbo, I., Jacobsson, K.G., van Doorn, C. and Petterson, G. (1992) A note on relations between oral stereotypies in dairy cows and milk production, health and age. *Animal Production* 54, 166–168.
Richardson, J.S. and Zaleski, W.A. (1986) Endogenous opioids and self-mutilation. *American Journal of Psychiatry* 143, 938–939.
Richardson, J.S., Holmlund, J.A., Gutkin, A., Blakemore, B. and Zaleski, W.A. (1986) On the role of endogenous opioids in the maintenance of self-mutilation. *International Journal of Neuroscience* 31, 129.
Ridley, R.M., Baker, H.F. and Haystead, T.A.J. (1981) Perseverative behaviour after amphetamine; dissociation of response tendency from reward association. *Psychopharmacology* 75, 283–286.
Schouten, W.G.P. and Wiepkema, P.R. (1991) Coping styles of tethered cows. *Behavioural Processes* 25, 125–132.

Singer, G.M., Wayner, M.J., Stein, J., Cimino, K. and King, K. (1974) Adjunctive behaviour induced by wheel running. *Physiology and Behaviour* 13, 493–495.

Sørensen, G. (1987) Animal experiments indicating behavioural pathologies as high cost strategy of survival. In: Checkland, P. and Kiss, I. (eds), *Problems of Constancy and Change. The Complementarity of Systems Approaches to Complexity*. International Society for General Systems Research, Hungary, pp. 1059–1063.

Sørensen, G. and Randrup, A. (1986) Possible protective value of severe psychopathology against lethal effects of an unfavourable milieu. *Stress Medicine* 2, 103–105.

Terlouw, E.M.C., Lawrence, A.B. and Illius, A.W. (1992) Relationship between amphetamine and environmentally induced stereotypies in pigs. *Pharmacology, Biochemistry and Behavior* 43, 347–355.

Valenstein, E.S. (1976) Stereotyped behaviour and stress. In: Serban, G. (ed.), *Psychopathology of Human Adaptation*. Plenum Press, New York and London, pp. 113–124.

Valzelli, L. (1980) *An Approach to Neuroanatomical and Neurochemical Psychophysiology*. Wright-PSG, Bristol.

van den Bos, R. and Cools, A.R. (1989) The involvement of the nucleus accumbens in the ability of rats to switch to cue-directed behaviours. *Life Sciences* 44, 1697–1704.

Watkins, L.R. and Mayer, D.J. (1982) Organization of endogenous opiate and non-opiate pain control systems. *Science* 216, 1185–1192.

Wayner, M.J. (1970) Motor control functions of the lateral hypothalamus and adjunctive behavior. *Physiology and Behavior* 5, 1319–1325.

Wayner, M.J. (1974) Specificity of behavioral regulation. *Physiology and Behavior* 12, 851–869.

Wiepkema, P.R. (1983) On the significance of ethological criteria for the assessment of animal welfare. In: Smidt, D. (ed.), *Indicators Relevant to Farm Animal Welfare*. Martinus Nijhoff, Boston, pp. 71–79.

Yates, A.J. (1962) *Frustration and Fixation. Frustration and Conflict*. Methuen, New York and London.

Yokel, R.A. and Wise, R.A. (1975) Increased lever pressing for amphetamine after pimozide in rats: implications for a dopamine theory of reward. *Science* 187, 547–549.

Conclusions and Implications for Animal Welfare

IAN J.H. DUNCAN[1], JEFFREY RUSHEN[2] AND ALISTAIR B. LAWRENCE[3]

[1]*Ontario Agricultural College, University of Guelph, Ontario, Canada:* [2]*Agriculture Canada Research Station, Lennoxville, Quebec, Canada:* [3]*The Scottish Agricultural College, Edinburgh, UK.*

Introduction

The study of stereotypic behaviour has provided useful information about the organization of behaviour (e.g. Fentress, 1973, 1976) and about the neurochemical basis of behaviour (Chapter 6). Interest in stereotypic behaviour, therefore, is not restricted to its implications for animal welfare. Nevertheless, much of the recent surge of interest in these behaviour patterns (Chapter 1) comes from their possible value as indicators of reduced welfare. Use of stereotypic behaviour to indicate problems in the environments of zoo animals has a long history (e.g. Hediger, 1934, 1950, 1955; Morris, 1964; Meyer-Holzapfel, 1968), which continues to the present (e.g. Kastelein and Wiepkema, 1989; Carlstead *et al.*, 1991). The use of stereotypic behaviour to assess the welfare of farm animals was formalized by Broom (1983) and Wiepkema *et al.* (1983) who claimed that the relative incidence of these behaviours indicates the relative state of welfare of animals in different environments.

It is frequently implied that the performance of stereotyped movements by itself is sufficient to indicate reduced welfare (e.g. Broom, 1986; Fraser and Broom, 1990; see also Chapter 1). However, there is a tendency towards circularity in this argument; a more stringent line of reasoning would attempt to define welfare independently and then look for the occurrence of stereotyped movements both when welfare was predicted to be good and when it was predicted to be poor. To date this has only been done in a rather haphazard way. In the previous chapters there have been many allusions to welfare, but now it is time to be explicit. What does the performance of stereotyped movements tell us about an animal's welfare? In order to deal with this properly, it is necessary to consider in some detail what is meant by 'welfare'.

What is Animal Welfare?

Since the publication of Ruth Harrison's book *Animal Machines* in 1964, there has been a burgeoning literature on animal welfare. Most people have fairly similar ideas about what is meant by 'animal welfare', and there would probably be general agreement that an animal obviously in pain, frightened or diseased would have poor welfare and that a healthy animal growing, reproducing and behaving 'normally' in a 'natural' environment would have good welfare. Between these two obvious extremes, however, there is a large grey area with lots of disagreement.

With the increased scientific interest in animal welfare, there have been numerous attempts to define exactly what was meant by the term. The committee that was formed by the British Government under the chairmanship of Professor Rogers Brambell to examine the welfare of animals in intensive housing systems (Command Paper 2836, 1965) decided that:

> Welfare is a wide term that embraces both the physical and mental well-being of the animal. Any attempt to evaluate welfare, therefore, must take into account the scientific evidence available concerning the feelings of animals that can be derived from their structure and functions and also from their behaviour.

In our opinion, the members of this committee got to the nub of the matter with their reference to 'feelings'; they were well ahead of their time in realizing that welfare has much to do with what animals feel. Duncan and Dawkins (1983) reviewed definitions and descriptions of welfare that appeared in the two decades following the Brambell Committee. They thought that none was perfect and concluded that it was impossible to give 'welfare' a precise scientific definition. They thought that a broad working definition would be one that included the ideas of: (i) the animal in physical and mental health; (ii) the animal in harmony with its environment; (iii) the animal being able to adapt to its environment without suffering; (iv) some account being taken of the animal's feelings. Similarly, they thought that a loose working definition of 'suffering' would be 'A wide range of unpleasant emotional states'.

At first, this suggestion of using broad working definitions of 'welfare' seemed generally acceptable to the scientific community investigating the topic. For example, Hughes (1976) defined welfare as 'a state of complete mental and physical health, where the animal is in harmony with its environment'. Hurnik et al. (1985) defined welfare as 'a state or condition of physical and psychological harmony between the organism and its surroundings'. However, as might be expected, the rapidly increasing number of investigations into welfare revealed exceptions which did not seem to fit these broad definitions. Many of these exceptions involved cases where the different components of welfare diverged. For example, what is the welfare status of an animal, otherwise normal, but with subclinical disease? What about an animal that is showing some symptoms of stress because of participation in a rewarding activity such as sexual activity (Szechtman et al., 1974; Colborn et al., 1991)? What of a healthy, physiologically normal animal that is performing stereotyped movements?

These 'problem cases' led to attempts to find a more precise definition of welfare. However, this led to divergence of opinion as to which aspect of animal welfare was the most fundamental. Duncan and Dawkins (1983), Dawkins (1990), Duncan and Petherick (1991) and Duncan (1993) argue that animal welfare is concerned partly or wholly with suffering that the animal consciously experiences. According to Duncan (1990), animal welfare is compromised only to the extent that animals actually suffer and as long as the mental state of the animals is protected then their welfare will be protected. Duncan (1990) distinguishes between 'needs', which are essential for survival and reproduction, and 'wants' which are the animal's cognitive representations of its needs. Welfare is primarily concerned with 'wants' not 'needs'. For example, regardless of whether or not pigs need straw bedding to maintain thermal equilibrium, if they want straw and are unable to obtain it then their welfare will be compromised. Wemelsfelder (Chapter 4) makes the similar claim that suffering (welfare) is essentially a concept dealing with subjectivity in animals.

This view focuses on animal subjectivity and emotions, which have long been neglected areas of biology (Rollin, 1989). It emphasizes that adequate nutrition and good health do not ensure good welfare, and that we must deal with the animals' perceptions of their environments not just with the environments themselves. Furthermore, recent experimental findings show that 'needs' can diverge from 'wants': despite decades of research into the nutritional requirements of gestating sows to ensure optimal reproduction, it is apparent that the amounts fed are insufficient to reduce feeding motivation (Lawrence et al., 1988) and that this may underlie stereotypic behaviours in sows (Chapter 3). To some, theories concerned with animals' feelings appear nebulous and difficult to probe experimentally. Nevertheless, this is the direction in which investigators appear to be heading (Duncan and Petherick 1989, 1991; Dawkins, 1990; Duncan, 1993). So far, researchers have stuck to relatively simple concepts such as fear or aversion that can be related closely to observable behaviour (Rushen, 1986). Wemelsfelder (Chapter 4), however, has attempted to develop a far more complex concept of animal subjectivity, which encompasses a wide range of mental states, such as boredom, and which is of relevance to stereotypic behaviour. An essential aspect of this concept is the animal's perception of the passage of time, which is regrettably overlooked in much behavioural research. Despite the difficulty in relating such complex ideas to empirical research, it is clear that long-term resolution of animal welfare problems requires us to pay more attention to understanding the psychological world of animals.

However, the conclusion that animal welfare is only about animal suffering has the danger that the debate about animal welfare will be seen as dealing only with imponderables (Fraser, 1989) and that attention will be drawn away from the problems of reproductive failures, disease and mortality that are still present in a large degree in many modern housing systems for confined animals (Fox, 1984; Ekesbo, 1988). Most veterinarians would feel that the welfare of an animal with subclinical disease was reduced, even if the disease is not yet perceptible to the animal. Other views on animal welfare emphasize these biological or health-related aspects of animal welfare. Broom (1991), for example, emphasizes biological fitness (i.e. longevity and reproductive success) as underlying good welfare. Hurnik (1988) suggests that welfare is concerned with the satisfaction of 'needs', which he defines as requirements for normal development and maintenance of good health. He distinguishes these from 'desires' (equivalent to Duncan's 'wants'), which are concerned with motivation and experience, and which he considers of lower priority for animal welfare. Curtis (1987) proposes a similar order of priority by suggesting that physiological needs are more important than behavioural needs. However, there are problems with this approach to welfare as well, particularly because some public concerns may be overlooked. Is it unreasonable to disregard intense pain or panic in animals even if this is sufficiently short-lasting as to have little effect on their biological functioning? Further, there are practical problems in placing different aspects of welfare in some order of priority. For example, does the slight reduction in disease incidence achieved by housing animals individually (e.g. Webster, 1991) justify the resulting social and behavioural deprivation that affect a much larger number of animals?

It sometimes seems as if these are two opposing and incompatible views on animal welfare: one emphasizing emotional suffering and the other emphasizing biological functioning. Unfortunately, defenders of each view often write as if these views were mutually exclusive. However, this is to ignore the evidence for reciprocal links between behavioural and physiological processes. For example, the work of Mason (1974) and others (see Chapter 5 and Dantzer and Mormède, 1983a) has illustrated the intricate interconnections between the emotional and cognitive responses of the animal to changes in a whole array of physiological variables. Furthermore, the established relationships between 'stress' and immune-competence (e.g. Tecoma and Huey, 1985), and the growing evidence for reciprocal links between products of the immune system (e.g. Interleukin 1) and behaviour (Tazi *et al.*, 1988), emphasize the close integration of behaviour, physiology and immunology. No doubt a concept will eventually arise that successfully integrates these aspects of welfare. In the meantime, to avoid semantic disputes and to aid progress in the area of animal welfare, these two views can be reconciled by the simple expedient of recognizing that there are two aspects of animal welfare, one concerned more with the immediate subjective experience of animals, and hence the extent of suffering, the other more concerned with the long-term biological functioning of animals. One aspect should not be emphasized at the expense of the other, and the established links between the two acknowledged. Likewise, Fraser (1989) recognizes the multifaceted nature of animal welfare, and points out the danger in mistaking one component of welfare for the whole.

On this basis we will consider the relationship between stereotypies and welfare with respect to the extent they indicate suffering or are associated with challenges to animals' biological functioning.

The Classification of Stereotypic Behaviour

Before dealing with the main issue of animal welfare, it is necessary to determine if the category of 'stereotyped behaviour' is a useful one. It is now apparent that the behaviour in the category of 'stereotyped movements' is not homogeneous with regard to form, causation or development (Chapter 2), motivation (Chapter 3), underlying physiology (Chapter 6) or function (Chapter 7).

Mason (1991, Chapter 2) has provided a valuable service in pointing out the evidence that the category of stereotypy is a heterogeneous one. In addition, there is often no clear demarcation between stereotypic behaviour and 'normal' behaviour. As is usual in such cases, differences between normal and stereotypic behaviour are often clear at the extremes but there is a large, intermediate area where the classification as normal or stereotypic is very difficult to make. Indeed, the first two properties in the now-classic definition of Ödberg (1978) (i.e. fixed form and repetitiveness), are properties of normal behaviour. This requires the inclusion of the third property of the definition, viz. the lack of function. As expected, this has caused a number of problems for some people. For example, Baxter (1989) feels that such a property is always open to challenge by those who feel that they have identified a function of stereotypies. Certainly, the 'lack of function' reflects more our present state of knowledge than any intrinsic property of the behaviour (see Chapter 2). However, this appears more serious than it is. As long as the aim of the definition is merely to identify particular behaviour as worthy of further investigation then the definition is quite adequate. However, we must realize that simply by naming a behaviour pattern as a 'stereotypy', we have not discovered anything about the behaviour. Most importantly, the lack of an identified function does not mean that we can assume that the behaviour reflects poor welfare.

A more serious problem is the methodological one. The lack of homogeneity of stereotypies and the fact that they form a continuum with normal behaviour means that we must give far more attention to describing in detail the particular behaviour that we are considering as stereotypies. This point is well made by Mason (Chapter 2) but it is worth extra emphasis. A detailed description of behaviour as well as the criteria used to distinguish stereotypies from other behaviour is an essential aspect of the scientific method and is essential to allow replication and comparison of different studies. The lack of such description is a common problem in applied ethology (Fraser and Rushen, 1987). In particular, we need good descriptions of the particular types of behaviour involved, measures of variability and degree of stereotypy, and descriptions of the circumstances under which the behaviour is performed. Researchers should avoid lumping behaviourally distinct forms of stereotypy. The lack of methodological rigour in the descriptions of stereotypies in most research papers is one of the main causes of the conflicting results that are often obtained.

Several of the accompanying chapters have discussed behaviours, which while not strictly stereotypic, appear closely related in some respects, such as adjunctive behaviour (e.g. schedule-induced polydipsia (SIP); see Chapter 7) and displacement activities (see Chapter 3). There is no doubt that these are interesting examples with the great advantage to researchers that they can be elicited predictably. Consequently, they have often served as 'models' of environmental stereotypies. However, whether or not they are analogous to the environmentally-induced stereotypies seen in zoo, farm and some laboratory animals remains uncertain. Some differences in underlying physiology have been pointed out in Chapters 5 and 7. Not only that, but the results from the SIP paradigm do not seem to apply to all adjunctive activities. For example, wheel-running in rats (Tazi *et al.*, 1986) and chain-chewing in pigs (Dantzer and Mormède, 1983b), both of which can be induced in a similar manner to SIP, can have different physiological correlates. There are also many differences between stereotypies and displacement activities. For example, compared to stereotyped movements, displacement activities: (i) occur frequently in natural situations (frequently enough to have been a great source of study for early ethologists); (ii) occupy a relatively small part of the animal's time; (iii) give no impression of being addictive; (iv) do not change with repeated performance. This is not to say that future studies of adjunctive behaviour or displacement activities will not continue to have use as models to answer some of the questions associated with stereotyped movements. However, at the moment they appear sufficiently different for more caution to be required in using data from them to draw direct conclusions about the welfare implications of stereotypies.

Stereotypies and Welfare

As has been pointed out clearly by Mason (1991), the question of the connection between the performance of stereotypies and welfare probably requires to be addressed at least at two stages of development; when the stereotyped movements are developing and once they are established as true stereotyped movements. Indeed, recognizing that the causes and consequences of stereotypies may change as they develop is essential for future research (Chapter 8).

This volume clearly illustrates that we have some way to go before we have a proper understanding of the relationship between stereotypies and suffering. Wemelsfelder (Chapter 4) most directly tackled this in her theoretical framework, by linking performance of stereotypies to the subjective experiences of animals, and specifically to boredom. In a human context, boredom is often used to describe a situation lacking (general) stimulation, and in this sense the idea of boredom as a cause of stereotypies may seem at odds with the evidence that stereotypies are significantly reduced when sows and broiler breeders are offered substantially more food (see above and Chapter 3). Wemelsfelder, however, uses animal boredom in a rather different sense to describe the impairment, by restrictive housing, of the animal's tendency to show spontaneous, voluntary interaction with the environment (i.e. to seek active control rather than simply to react to stimulation). She argues that active control signifies the subjective nature of animal behaviour, and the impairment of this underlies stereotyped behaviour. Thus, according to Wemelsfelder, both developing and established stereotypies signify suffering. The value of this theoretical approach now requires to be tested empirically.

Other chapters have dealt less explicitly with stereotypies and suffering, but have been concerned much more with the behavioural and physiological mechanisms underlying the behaviour. Broadly we can identify three current approaches:

1. The first approach assumes that stereotypies are a non-specific response to aversive emotional states such as frustration, and/or high levels of arousal. It has often been suggested that during their development, stereotypies arise in response to the aversive (e.g. frustrating) aspects of intensive housing (see Chapters 3 and 8). Although largely circumstantial, this view is supported by the observation that stereotyped movements can be induced under conditions of apparent frustration or conflict, which are likely to be associated with aversion (see e.g. Duncan and Wood-Gush, 1972; also Gray, 1987). In a more mechanistic approach Dantzer (1986) has suggested that stereotypies are the expression of the arousal engendered by aversive aspects of the environment (see also Fentress, 1976).

2. The second approach relates stereotypies to specific motivational states that control normal behaviour without immediately considering the mental

state of the stereotyping animal. It has become increasingly apparent that processes involved in the control of 'normal' behaviour may also underlie stereotypies (Chapter 3). For example, there is now substantial evidence that stereotypies in sows and broiler breeders are very closely related to food restriction and resulting hunger, and there is little evidence to link such stereotypies with aversion (Chapter 3) or stress (Chapter 5), although in many cases this may well reflect on the lack of research. Such results can be used to generate models of stereotypies based on motivational processes such as positive feedback, that leave open the issue of whether suffering is involved (see also Chapter 4). As an example, Hughes and Duncan (1988) propose a behavioural model in which stereotypies result from the continued performance of appetitive sequences of behaviour; this approach makes little reference to any negative emotional consequences of this process.

3. A mixture of approaches 1 and 2. At this point in time we think it most likely that the development of stereotypies depends on a complex interaction between a specific motivational state(s), and progressive effects of arousal/stress (as a response to aversive mental states such as frustration) on the neurochemical control of behavioural output. For example, although stereotypies may be an expression of a specific motivational state (e.g. as suggested by Hughes and Duncan, 1988), aversion, by increasing arousal, may contribute to the persistence of the behaviour by reducing the tendency to rest (see Chapter 3). More work on the neurochemical basis of stereotypies is likely to reveal further links between motivation, stress and suffering. Cabib (Chapter 6) sees 'stress' arising when the animal suffers 'loss of control' (viz. Weiss et al., 1981) over an arousing stimulus. Stress, in this sense, may facilitate performance of a motivated behaviour as stereotyped sequences, through inducing subtle changes in the self regulatory processes of the brain's dopamine systems (see Chapter 6). Future research could consider if Wemelsfelder's 'impairment of active control', is in any way equivalent in its effects to the 'loss of control' induced in the classic experiments of Weiss and others, and which Cabib suggests may be fundamental to the neurochemical changes accompanying the development of stereotypies (Chapter 6).

There also remains uncertainty over the relationship between suffering and established stereotypies. In contrast to Wemelsfelder (Chapter 4; see above), Dantzer (1986) suggests that established stereotypies are controlled at such a 'hard-wired' level, that they cease to have any emotional significance to the animal (see also Chapter 8). Again these ideas require rigorous empirical analysis of their predictions. One experimental approach that might be considered is the use of choice tests, to ask animals to choose between environments to indicate how they 'feel' about the conditions under which they are kept and the procedures to which they are subjected (Duncan, 1993; see also Cooper and Nicol, 1991). But here is the stumbling block; the state of animals performing stereotyped movements may prevent them from answering the question, if stereotypies are indeed the outward expression of disturbances in brain neural structures governing the persistence of behaviour (see above; Chapter 6; Dantzer, 1991). Indeed, much of the interest in the neurochemical bases of these behaviours comes from the possibility that they may serve as models for human psychopathology (Chapter 6). In this respect, stereotyping animals may be similar to human beings with psychological pathologies. Just as it may be unrealistic to ask human beings with such pathologies how they feel, so we cannot place any reliance on the answers we might get from animals in these situations. Wemelsfelder (Chapter 4) makes a similar point.

Next, we will consider whether stereotypies indicate that the biological functioning of animals is impaired. Given the important reciprocal relationships between behaviour, physiology and immunology we are partly considering here the translation of the emotional (or subjective) state of the stereotyping animal into biological correlates. This research generally involves looking for correlations between the performance of stereotypies and physiological measures of 'stress'. The results of this research are summarized in Chapters 5, 6 and 7, and the following points emerge:

1. While the occurrence of some stereotypies is associated with increases in pituitary–adrenocortical or sympathetic-nervous activity, no general conclusions can be drawn that are valid for all stereotypies.
2. Present concepts of stress, particularly those that equate stress with measures of corticosteroids, are too simplistic to capture the complex effects of the environment upon animal physiology (Chapter 7). We need to develop more complex ideas and a greater appreciation of the value of interdisciplinary work (Chapter 8). In particular, more attention needs to be paid to events in the central nervous system rather than peripheral physiological systems.
3. It is difficult to infer much about the causal relationships between the performance of stereotypies and physiology from correlation studies alone. In particular, there is a need for more experimental research to separate out the physiological consequences of stereotypies from physiological differences that might predispose animals to develop stereotypies. There is as yet no unequivocal evidence that stereotypies themselves affect an animal's physiology.
4. The simple hypothesis that stereotypies might be a coping response must be discarded (see Chapters 5 and 7). The suggestion of Cronin (1985) that stereotyping pigs might be narcotizing themselves, and thus reducing physiological responses to stress, seems to have been rebutted (Chapters 5 and 7). The relationship between endogenous opioid peptides and the performance of stereotypies is complex (Chapters 5 and 6). The tendency of applied ethologists to favour simplistic models of this relationship has been heavily criticized (Dantzer, 1991, Chapter 7). This may reflect their failure to appreciate older work or a lack of acquaintance with work in other disciplines (Chapter 8). This leaves the possibility open that the performance of stereotyped

movements may have a coping component (but in a much more complex interaction than the simple release of endogenous opioids; see Chapters 7 and 8).

Last, we should consider how we can apply the information assembled in this book to the practicalities of housing animals. Clearly the lack of certainty over the relationship between stereotypies and suffering (see above) is a major obstacle to using the incidence of stereotypies as an aid to future housing designs. We feel on balance, however, that this position reflects on the lack of research (see Chapter 1), and on the difficulties of studying events in the central nervous system in large animals (see above). On this basis we recommend that animals be 'given the benefit of the doubt', and until examples are produced of stereotypies that clearly develop without a strong negative emotional reaction, it is safer to assume that, at least in their development, they indicate some measure of arousal, stress and consequently suffering.

On the basis that stereotypies may reflect past or present suffering we should consider how to design housing to prevent occurrence of this behaviour. The following approaches should be considered:

1. Reduction in the underlying motivational state: This unfortunately may often not be as straightforward as it at first appears. For example, increasing food allowances to food-restricted animals may lead to obesity and increased costs of production, although certain 'bulky' foods, such as sugarbeet pulp, may temporarily reduce feeding motivation due to their physical properties, and be used to provide an *ad libitum* food source that still prevents undue nutrient intake (e.g. Brouns *et al.*, 1991). It may also be difficult for a number of reasons to identify a specific underlying motivational cause of a stereotypy; for example, channelling (see Chapter 3) could markedly modify the final behavioural expression of the state. Last, it will also be important to consider other aspects of the environment, such as the behaviour of stockpersons, that are not directly related to the underlying motivation but that increase arousal/stress (see above). For example, Appleby *et al.* (1989) found evidence in sows that the noise from chain-chewing neighbours was additive to the effect of food restriction on performance of stereotypy.

2. Expression of species-typical behaviour: There has often been a strong case made by certain applied ethologists (and also the public) that confined animals should be allowed to perform 'natural' behaviour. This position has been attacked on the grounds that it is often based on the concept of 'Lorenzian behavioural needs' (see Chapter 3; e.g. Dawkins, 1983). This volume has, however, produced new grounds to support allowing expression of some species-typical behaviour as a means of reducing the incidence of stereotypies. First, given that it may be difficult to reduce the underlying motivational state (see above), the logical alternative is to allow that state to be expressed in such a way that the behavioural output is not channelled by the restrictive environment into simple and often repeated elements (Chapter 3). The static and

unvarying nature of intensive environments appears to be the essential characteristic that leads to channelling (see Chapter 3), and therefore animals experiencing high levels of a motivational state (or arousal) should be housed in environments that give the opportunity to express that state in a suitably complex form. Second, allowing motivated behaviour to be expressed in a more species-typical form may be the only means of preventing frustration where reduction in the underlying motivation cannot be achieved. It could also be construed as giving back active control to animals, and thereby preventing the process of boredom described by Wemelsfelder (Chapter 4).

Conclusions

The problems outlined above mean that it is unlikely that we can use variation in the incidence of stereotypies to measure welfare in any simple fashion. Even if we cannot yet draw firm conclusions about how an animal performing stereotyped movements actually feels or how its physiology is affected, we would suggest that the following points are considered:

1. For some stereotypies, there is evidence that the animals have been through a period involving negative feelings such as aversion.
2. Some forms of stereotypy appear symptomatic of a psychopathology.
3. There is some evidence from the underlying physiology that some stereotyping animals are stressed.
4. Animals performing stereotypies may have negative feelings associated with boredom (or reduced active control).
5. Stereotypies are not simple coping responses.
6. The lack of evidence of a relationship between most stereotypies and reduced welfare reflects a lack of research rather than negative findings.
7. Enough is known about the eliciting environmental conditions to prevent the occurrence of some stereotypies.
8. We suggest for the reasons given above, that suitable provision for species-typical behaviour will in many cases largely eradicate performance of stereotypies.

Given these points, we recommend that appropriate steps are taken to prevent the occurrence of stereotypies in the name of animal welfare.

References

Appleby, M.C., Lawrence, A.B. and Illius A.W. (1989) Influence of neighbours on stereotypic behaviour in tethered gilts. *Applied Animal Behaviour Science* 24, 137–146.

Baxter, M. (1989) Philosophical problems underlying the concept of welfare. In:

Faure, J.M. and Mills, A.D. (eds), *Third European Symposium on Poultry Welfare*. WPSA, Tours, pp. 59–66.

Broom, D.M. (1983) Stereotypies as animal welfare indicators. In: Smidt, D. (ed.), *Indicators Relevant to Farm Animal Welfare*. Martinus Nijhoff, The Hague, pp. 81–87.

Broom, D.M. (1986) Indicators of poor welfare. *British Veterinary Journal* 142, 524–526.

Broom, D.M. (1991) Assessing welfare and suffering. *Behavioural Processes* 25, 117–123.

Brouns, F., Edwards, S.A., English, P.R. and Taylor, A.G. (1991) Effects of diet and feeding regime on behaviour of group housed pregnant gilts. In: Appleby, M.C., Horrell, R.I., Petherick, J.C. and Rutter, S.M. (eds), *Applied Animal Behaviour: Past, Present and Future*. UFAW, Potters Bar, UK, pp. 143–144.

Carlstead, K., Seidensticker, J. and Baldwin, R. (1991) Environmental enrichment for zoo bears. *Zoo Biology* 10, 3–16.

Colborn, D.R., Thompson, D.L., Roth, T.L., Capehart, J.S. and White, K.L. (1991) Responses of cortisol and prolactin to sexual excitement and stress in stallions and geldings. *Journal of Animal Science* 69, 2556–2562.

Command Paper 2836 (1965) *Report of the Technical Committee to Enquire into the Welfare of Animals Kept under Intensive Husbandry Systems*. HMSO, London.

Cooper, J.J. and Nicol, C.J. (1991) Stereotypic behaviour affects environmental preference in bank voles (*Clethrionomys glareolus*). *Animal Behaviour* 41, 971–977.

Cronin, G.M. (1985) The development and significance of abnormal stereotyped behaviours in tethered sows. PhD Thesis, Agricultural University of Wageningen, The Netherlands.

Curtis, S.E. (1987) Animal well-being and animal care. *Veterinary Clinics of North America: Food Animal Practice* 3, 369–382.

Dantzer, R. (1986) Behavioural, physiological and functional aspects of stereotyped behaviour: a review and reinterpretation. *Journal of Animal Science* 62, 1776–1786.

Dantzer, R. (1991) Stress, stereotypies and welfare. *Behavioural Processes* 25, 95–102.

Dantzer, R. and Morméde, P. (1983a) Stress in farm animals: a need for a re-evaluation. *Journal of Animal Science* 57, 6–18.

Dantzer, R. and Morméde, P. (1983b) Dearousal properties of stereotyped behaviour: evidence from pituitary–adrenal correlates in pigs. *Applied Animal Behaviour Science* 10, 233–244.

Dawkins, M.S. (1983) Battery hens name their price: consumer demand theory and the measurement of ethological 'needs'. *Animal Behaviour* 31, 1195–1205.

Dawkins, M.S. (1990) From an animal's point of view: motivation, fitness, and animal welfare. *Behavioural and Brain Sciences* 13, 1–9.

Duncan, I.J.H. (1990) Animal welfare: what is it and how can we measure it? Presented to Alberta Institute of Agrologists, Edmonton, 19 October, 1990.

Duncan, I.J.H. (1993) Welfare is to do with what animals feel. *Journal of Agriculture and Environmental Ethics* (in press).

Duncan, I.J.H. and Dawkins, M.S. (1983) The problem of assessing 'well-being' and 'suffering' in farm animals. In: Smidt, D. (ed.), *Indicators Relevant to Farm Animal Welfare*. Martinus Nijhoff, The Hague, pp. 13–24.

Duncan, I.J.H. and Petherick, J.C. (1989) Cognition: the implications for animal welfare. *Applied Animal Behaviour Science* 24, 81.

Duncan, I.J.H. and Petherick, J.C. (1991) The implications of cognitive processes for animal welfare. *Journal of Animal Science* 69, 5017–5022.

Duncan, I.J.H. and Wood-Gush, D.G.M. (1972) Thwarting of feeding behaviour in the domestic fowl. *Animal Behaviour* 20, 444–451.

Ekesbo, I. (1988) Animal health implications as a result of future livestock and husbandry developments. *Applied Animal Behaviour Science* 20, 95–104.

Fentress, J.C. (1973) Specific and non-specific factors in the causation of behaviour. In: Bateson, P.P.G. and Klopfer, P. (eds), *Perspectives in Ethology*. Plenum Press, New York, pp. 155–224.

Fentress, J.C. (1976) Dynamic boundaries of patterned behaviour: interaction and self-organization. In: Bateson, P.P.G. and Hinde, R.A. (eds), *Growing Points in Ethology*. Cambridge University Press, Cambridge, pp. 135–167.

Fox, M.F. (1984) *Farm Animals: Husbandry, Behaviour and Veterinary Practice*. University Park Press, Baltimore.

Fraser, A.F. (1989) Animal welfare practice: primary factors and objectives. *Applied Animal Behaviour Science* 22, 159–176.

Fraser, A.F. and Broom, D.M. (1990) *Farm Animal Behaviour and Welfare*. Baillière Tindall, London.

Fraser, D. and Rushen, J. (1987) A plea for precision in describing ethological methods. *Applied Animal Behaviour Science* 18, 205–209.

Gray, J.A. (1987) *The Psychology of Fear and Stress*. Cambridge University Press, Cambridge.

Harrison, R. (1964) *Animal Machines*. Vincent Stuart, London.

Hediger, H. (1934) Uber bewegungstereotypien bein gehaltenen Tieren. *Revue Suisse Zoologie* 41, 349–356.

Hediger, H. (1950) *Wild Animals in Captivity*. Butterworth, London.

Hediger, H. (1955) *Studies of the Psychology and Behaviour of Captive Animals in Zoos and Circuses*. Butterworth, London.

Hughes, B.O. (1976) Behaviour as an index of welfare. In: *Proceedings of the Fifth European Poultry Conference, Malta*, pp. 1005–1018.

Hughes, B.O. and Duncan, I.J.H. (1988) The notion of ethological 'need', models of motivation, and animal welfare. *Animal Behaviour* 21, 10–17.

Hurnik, J.F. (1988) Welfare of farm animals. *Applied Animal Behaviour Science* 20, 105–117.

Hurnik, J.F., Webster, A.B. and Siegel, P.B. (1985) *Dictionary of Farm Animal Behaviour*. University of Guelph, Guelph.

Kastelein, R.A. and Wiepkema, P.R. (1989) A digging trough as occupational therapy for Pacific walruses (*Odobenus rosmarus divergens*) in human care. *Aquatic Mammals* 15, 9–17.

Lawrence, A.B., Appleby, M.C. and Macleod, H.A. (1988) Measuring hunger in the pig using operant conditioning: the effect of food restriction. *Animal Production* 47, 131–137.

Mason, J.W. (1974) Specificity in the organization of neuroendocrine response profiles. In: Seeman, P. and Brown, G. (eds), *Frontiers in Neurology and Neuroscience Research*. University of Toronto, Toronto, pp. 68–80.

Mason, G.J. (1991) Stereotypies and suffering. *Behavioural Processes* 25, 103–115.

Meyer-Holzapfel, M. (1968) Abnormal behaviour in zoo animals. In: Fox, M.W. (ed.), *Abnormal Behaviour in Animals*. Saunders, London, pp. 476–503.

Morris, D. (1964) The response of animals to a restricted environment. *Symposium of the Zoological Society of London* 13, 99–118.

Ödberg, F.O. (1978) Abnormal behaviours (stereotypies). In: *Proceedings of the 1st World Congress on Ethology Applied to Zootechnics*. Industrias Graficas Espana, Madrid, pp. 475–480.

Rollin, B. (1989) *The Unheeded Cry: Animal Consciousness, Animal Pain and Science*. Oxford University Press, New York.

Rushen, J. (1986) The validity of behavioural measures of aversion. A review. *Applied Animal Behaviour Science* 16, 309–323.

Szechtman, H., Lambrou, P.J., Caggiula, A.R. and Redgate, E.S. (1974) Plasma corticosterone levels during sexual behaviour in male rats. *Hormones and Behaviour* 5, 191–200.

Tazi, A., Dantzer, R., Mormède, P. and Le Moal, M. (1986) Pituitary–adrenal correlates of schedule-induced polydipsia and wheel-running in rats. *Behavioural Brain Research* 19, 249–256.

Tazi, A., Dantzer, R., Crestani, F. and Le Moal, M. (1988) Interleukin-1 induces conditioned taste aversion in rats: a possible explanation for its pituitary–adrenal stimulating activity. *Brain Research* 473, 369–371.

Tecoma, E.S. and Huey, L.Y. (1985) Psychic distress and the immune response. *Life Sciences* 36, 1799–1812.

Webster, A.J.F. (1991) Control of infectious diseases in housed veal calves. In: Metz, J.H.M. and Groenenstein, C.M. (eds), *New Trends in Veal Calf Production*. Pudoc, Wageningen, pp. 103–112.

Weiss, J.M., Goodman, P.A., Losito, B.G., Corrigan, S., Charry, J.M. and Bailey, W.H. (1981) Behavioural depression produced by an uncontrollable stressor: relationship to norepinephrine, dopamine, and serotonin levels in various regions of rat brain. *Brain Research Review* 3, 167–205.

Wells, M.J. (1962) *Brain and Behaviour in Cephalopods*. Heinemann, London.

Wiepkema, P.R., Broom, D.M., Duncan, I.J.H. and van Putten, G. (1983) Abnormal behaviour in farm animals. *Report of the Commission of the European Community, Luxembourg*, pp. 1–16.

Index

ACTH 104
  grooming induced by 126
  secretion 105–6
action-specific energy 48
acute stressors 100
adaptation 78
  to stressors 101
adjunctive behaviour 109–10, 179, 198
  and coping 152–8
adrenal function test 102
adrenalectomy, effect on SIP 155–6
adrenaline 99, 104
  secretion 106
age of subject, and form of
    stereotypy 12–13
albumin 106
alertness 76, 77
amphetamine 19, 121, 122–5, 158
  grooming induced by 126
  and self-organization 56
  sensitization to 130–1
  as a stressor 158–60
  see also stimulant-induced
    stereotypies
amygdala 102–4
animal subjectivity 67, 195–6

as capacity to interact with the
    environment 68–71
  criteria for measurement 71–7
  defining 68–77
animal welfare 2
  definitions 194–7
  and stereotypies 199–203
ANS 110–11
anxiety 83, 85–7, 183
apomorphine 19, 125, 158
appetitive behaviour 49, 165–6
arousal
  definition 129, 149
  effect of stereotypies on 178
  heterogeneity 31
  role in stereotypies 54–5, 111, 183
arrow of time 70
attentiveness 73
autism 23–4
autonomic nervous system 110–11
auto-shape 49
aversion 45–6

β-endorphin 105
behaviour
  environmental channelling 50–1

208 Index

behaviour *cont.*
  models of 66–7, 68–71
  motivational analysis 42
  normal, incorporation of stereotypies 10
  patterns
    inappropriate 10–11
    local rate 29
    transition from active to passive 66–7
    unvarying, development 8–10
behavioural
  arousal 55
  deprivation 47
  flexibility 72
  inhibition system 185–6
  innovation, and play 76
  need 47
  sensitization 101, 121, 130–1, 151
    and adaptation to environmental constraints 134–40
    and behavioural dysfunctions 138–40
    to stress 135–8
behaviourism, and animal subjectivity 68–9
beneficial stereotypies 25
BIS 185–6
blunting 149
body-rocking 16–17
  inappropriate 10
  in mentally handicapped children 9
body-shakes 125, 126
body temperature 162–3
boredom 52, 66, 83, 84–5, 199
brain
  damage 22–3
  plasticity 75–6
  stem 110
Brambell Committee 1, 2, 194

cage-induced stereotypies 15
calcium deficiency 44
carazolol 111
Cartesian dualism, and animal subjectivity 68, 69–70
caudatus putamen 123–4
causation 176
central nervous system, interface with endocrine systems 103

chain manipulation 109–10, 157, 198
changes in stereotypies with age 17–18
cholecystokinin 112
chronic intermittent stressors 100
chronic stressors 100
classification of stereotypies 7–31, 197–8
Clethrionomys britannicus 17
climbing response 135–6
cognitive mediator concept 99, 102
cognitivism, and animal subjectivity 69
competition 51
compulsive behaviour 24
concentration 76
confinement, effects of 1–2
conflict 53–4
conscious attention 76
consciousness 73
conservation-withdrawal state 185
consummatory behaviour 165, 166
control 67
control system models 52
coping strategies 78, 129, 148–50, 183
  active/passive 184
  and adjunctive activities 152–8
  and stereotypies 164–6
  and stimulant-induced stereotypies 158–62
coping–predictability concept 99, 102
corticosteroid binding globulin 106
corticosterone 135, 139–40, 183
  plasma levels
    in footshock 159
    induced by amphetamine injection 160–2
    in schedule-induced activities 154–8
corticotropic releasing hormone 104, 105–6
cortisol 109–10, 183
  plasma levels in chain-manipulation 157
CRH 104, 105–6
crib-biting 25

DA *see* dopamine
de-arousal 78

decortication 124
definitions 8, 10–11, 148, 175
degree of stereotypy 57
denial 149
depression 83, 85–7
deprivation stereotypies 12
desires 196
deterministic chaos 186–7
Deutung 174
development of stereotypies 8–10, 182
digestive function 112
disinhibition 51, 54
displacement behaviour 53–4, 139, 153, 198
dithering 52
dopamine (DA) 122–5, 126–8
  and behavioural adaptation to stress 135–8
  increased release and stereotypies 138–40
  responses to stress and stimulants 131–4
drugs, stimulant *see* stimulant-induced stereotypies
dynorphin 105

effort 76
emancipation 17
enkephalins 105, 127
environment
  adaptation to constraints in 134–40
  changes in 51–2
  design of 202–3
  enrichment 46, 80–1, 182
  historical research 176
  impoverishment 66, 79, 80, 82–3
  interaction with 66
    and animal subjectivity 68–71
    impaired capacity 80–3
  monotonous 87
  novel 124
  stereotypies induced by 11–18, 129–34
escape behaviour 153, 179
established stereotypies 17–18
ethological techniques 26–7
  relevance to pharmacology 177

etho-pharmacology 177
euphoria 164
experience, and level of stereotypy 14
exploration 71, 74–6, 81–2
exploratory motivation 46

feed-forward 71
feeding
  behaviour 15, 43–4, 56, 153
  and timing of stereotypies 49–50
feelings 194
fight and flight reaction 99
fimbria fornix, lesions 139
fitness 196
fixation 179
food
  bulk 43–4
  restriction 43
footshock, effect on plasma corticosterone levels 159
foraging behaviour 15, 43–4, 49
forepaw treading 125, 126
frontal cortex, lesions 125
frontal lobotomy 22–3
frustration 83–4, 183, 199
future-directedness 70

gastrin 112
general adaptation syndrome 99
genotype, and behavioural adaptation to stress 135–8
glucocorticosteroids 99
  secretion 105–6
goal-directed behaviour 165–6
grooming, excessive 126–7

habits 10, 79
habituation, to stressors 101
haloperidol 125
harmful stereotypies 24–5
head-weaving 125, 126
heart rate 111, 163
heterogeneity of stereotypies 11–26, 197–8
hind limb abduction 125, 126

hippocampectomy 23
hippocampus 125, 186
  lesions 124
hour-glass model of stress 99
HPA axis 104, 108–10
  in SIP 155–6
5-HT 125–6
humans, association of stereotypies with clinical conditions 23–4
6-hydroxydopamine 122, 132
5-hydroxytryptamine 125–6
hypothalamic–pituitary–adrenal axis *see* HPA axis
hypothalamus 102–4

identification of stereotypies 8–11
immunocompetence 196–7
incomplete stereotypies 18
individual differences
  in performance of stereotypies 25–6, 55–6
  in reactions to stress 101–2, 164–5
inquisitive exploration 74–5
insulin 112
interdisciplinarity 177, 178–80
interim stereotypies 16
interpretation of stereotyped behaviour 78–87
interstressor interval 101
isolation syndrome 178
Istwert 53, 181

L-dopa 22
learned helplessness 86
learning
  behavioural 71–2
    importance of orienting behaviour 73
  disabilities 23–4
  and play 76
leg-swinging 163
limbic system 102–4, 110
literature survey 3–4
locomotory behaviour 44–5
loose stereotypies 27
Lorenzian motivational model 47–8

Maier's paradigm 179
means-end-readiness 73
mechanistic models, and animal subjectivity 70–1
mesoaccumbens DA system 123–4, 126
  in behavioural adaptation to stress 135–8
  hyperactivity and stereotypies 138–40
  in response to stress and stimulants 130–4
mesocortical DA system 123, 131–2
mesolimbic DA system 123
Microtus agrestis 17
misbehaviour 49, 50
models 177, 198
  of behaviour 66–7, 68–71
  control system 52
  explanatory limitations 78–9
  motivational 42–3, 47–8
  of schizophrenia 121
modulating factors 181–2
monotony 87
morphine 22, 127
morphology 29–30
motivation 42–3, 56–8, 175
  flexibility of systems 49–51
  internal sources 47–9
  processes relevant to stereotypies 47–56
  specific systems underlying stereotypies 43–7
  transition between systems 51–2

naloxone 107, 108, 111, 163–4, 185
needs 195, 196
negative feedback 52–3
neocortex 103, 104, 110, 125
nervous energy 47–8
nest-building behaviour 15, 46
neurobiological research 120–1
neuropeptides 126–8
nigrostriatal DA system 123–4, 126, 131–2
noradrenaline 104
  secretion 106
nucleus accumbens 122–5, 186

nucleus solitarius 110
nutrient deficits 44

obsessive behaviour 24
6-OHDA 122, 132
opiod antagonists 107, 108, 111, 163–4, 185
opioid peptides 22, 99, 121, 185
  in modulation of autonomic activity 110–11
  responses to 127–8
  sensitization to 132
  and stereotypies 107–8, 163–4
  in stress reactions 104–5
orienting behaviour 71, 72–4, 81–2
  neurophysiological correlates 76–7
origins of stereotypies 57
oxygen consumption 163
oxytocin 104

pacifier-sucking 163
pain sensitivity 164
persistence of stereotypies 57–8
phenomenological models, and animal subjectivity 70
physical time 70
play 31, 71, 74–6, 81–2
position-stereotypy 179
positive feedback 52–3
prelimbic cortex 186
problem-solving, and orienting behaviour 73
propranolol 163
psychiatry 178–9
psychological time 70–1, 72, 86
psychostimulants, neural responses to 131–4

rating scales 29–30
reafference 71
rebound 48
reinforcement 68
repetitiveness 10–11, 182–5
  and development of stereotypies 9
  quantification 27–8

reproductive function 112
research
  historical perspective 174–80
  methods 26–30
  suggestions for future 187
rhythmical behaviour 73–4
rhythmical stereotypies 74
rigidity 10–11
  quantification 27–8
routines 79
running fit syndrome 127

schedule-induced polydipsia 16, 152–6, 157–8, 183
schedule-induced stereotypies 16, 17, 109–10, 152–8, 183
schizophrenia 23–4
  models of 121
self-directed stereotypies 85
self-mutilation 31
self-narcotization 164, 185
self-organization 42, 56, 78, 80
sensitization *see* behavioural sensitization
sensory deprivation 151
sexual motivation 46
sham activities 47–8
shaping 179
SIP 16, 152–6, 157–8, 183
Sollwert 53, 181
somatostatin 112
sows, tethering 45
space allowance 44
species differences in stereotypy development 16–17
stimulant-induced stereotypies 19–22, 122–8
  and coping 158–62
stimulus-change 74
stimulus-complexity 74
stimulus novelty 74
stress 2, 121, 129
  behavioural adaptation to 135–8
  and behavioural dysfunctions 138–40
  and behavioural stereotypies 151–2
  concept of 99

stress *contd.*
  hour-glass model 99
  neural responses to 131–4
  physiological measurements in research of 98
  reactions to 102–6, 129–30, 150, 200
stress signal 105, 106
stressors 99–101, 129–30, 151–2
  amphetamine as 158–60
  standard 102
striatum 177
suffering 199
  chronic 79, 83–7
  definition 194
  and established stereotypies 200–1
superstitious behaviours 179
symbol-stereotypy 179
synchronicity 71

tail-pinch 151
temporal patterns 30
terminal stereotypies 15
terminology 180
  inadequacy 175–6
tethering, sows 45
thumb-sucking, inappropriate 10

tics 24
time 70–1, 72, 86
Tourette's syndrome 24
transcortin 106

unifying theoretical framework 180–7

vacuum activities 47–8
variability, quantification 27–8
vasopressin 104
ventral tegmental area 123
vice 85
volition 72, 76
voluntary attention 76

wants 195
welfare *see* animal welfare
well-being 67–8
wheel-running, schedule-induced 157, 198
wind-sucking 24–5